Chiral
Symmetry in
Hadrons and Nuclei

Chiral
Symmetry in
Hadrons and Nuclei

Editors

Li-Sheng Geng
Beihang University, China

Jie Meng
Peking University, China

Qiang Zhao
Institute of High Energy Physics, CAS, China

Bing-Song Zou
Institute of Theoretical Physics, CAS, China

World Scientific

NEW JERSEY · LONDON · SINGAPORE · BEIJING · SHANGHAI · HONG KONG · TAIPEI · CHENNAI

Published by

World Scientific Publishing Co. Pte. Ltd.
5 Toh Tuck Link, Singapore 596224
USA office: 27 Warren Street, Suite 401-402, Hackensack, NJ 07601
UK office: 57 Shelton Street, Covent Garden, London WC2H 9HE

British Library Cataloguing-in-Publication Data
A catalogue record for this book is available from the British Library.

CHIRAL SYMMETRY IN HADRONS AND NUCLEI
Proceedings of the Seventh International Symposium

ISBN 978-981-4618-21-2

Printed in Singapore

Contents

Preface   xi

Studies of chirality in the MASS 80, 100 and 190 regions
    R. A. Bark, E. O. Lieder, R. M. Lieder, E. A. Lawrie,
    J. J. Lawrie, S. P. Bvumbi, N. Y. Kheswa, S. S. Ntshangase,
    T. E. Madiba, P. L. Masiteng, S. M. Mullins, S. Murray,
    P. Papka, O. Shirinda, Q. B. Chen, S. Q. Zhang, Z. H. Zhang,
    P. W. Zhao, C. Xu, J. Meng, D. G. Roux, Z. P. Li, J. Peng,
    B. Qi, S. Y. Wang and Z. G. Xiao   1

$\rho K\bar{K}$ system within the framework of the fixed center
approximation to Faddeev equations
    M. Bayar, W. Liang, T. Uchino and C. W. Xiao   9

Symmetry breaking in the parton distribution functions
of the nucleon
    F.-G. Cao   14

The predictions of $\psi(nS)$ and $\Upsilon(nS)$ decays based on the
molecular structure
    L. R. Dai and E. Oset   18

How to distinguish a molecule from an 'elementary' particle?
    L. Y. Dai, M. Shi, G. Y. Tang and H. Q. Zheng   22

A chiral effective approach to parity-odd proton-proton scattering
    J. de Vries   29

$\bar{K}N$ reactions and baryon resonances with strangeness -1
    Z.-H. Guo and J. A. Oller   33

Tensor force and Delta excitation for the structure of light nuclei
K. Horii, H. Toki and T. Myo   37

Compositeness of hadron resonances in chiral dynamics
T. Hyodo   41

Thermodynamics of hadrons using the Gaussian functional
method in the linear sigma model
S. Imai, H.-X. Chen, H. Toki and L.-S. Geng   49

Studies of hypernuclei with the AMD method
M. Isaka   53

Light quark mass dependence of the $X(3872)$ in XEFT
M. Jansen, H.-W. Hammer and Y. Jia   57

Hyperfine structure of ground-state nucleons in chiral quark model
D. Jia, W.-B. Dang and X.-W. Zhao   61

The high order chiral Lagrangian
S.-Z. Jiang and Q. Wang   70

The $f_0(1790)$ and $f_0(1800)$ puzzle
K. P. Khemchandani, A. Martínez Torres, M. Nielsen,
F. S. Navarra, D. Jido, A. Hosaka and E. Oset   74

Ab initio no core full configuration approach for light nuclei
Y. Kim, I. J. Shin, P. Maris, J. P. Vary, C. Forssén and
J. Rotureau   78

Understanding nuclear shape phase transitions within SD-pair
shell model
L. Li, Y. Zhang, X. Yuan, J. Li, Y. Luo, F. Pan and
J. P. Draayer   86

The $\eta K\bar{K}$ and $\eta'K\bar{K}$ systems with the fixed center
approximation to Faddeev equations
W.-H. Liang, C. W. Xiao and E. Oset   99

Heavy quark spin structure in hidden charm molecules
   Y.-R. Liu   103

Search for deeply bound Kaonic nuclear states via
^{3}He(K^-, n) reaction at J-PARC
   Y. Ma, S. Ajimura, G. Beer, H. Bhang, M. Bragadireanu,
   P. Buehler, L. Busso, M. Cargnelli, S. Choi, C. Curceanu,
   S. Enomoto, D. Faso, H. Fujioka, Y. Fujiwara, T. Fukuda,
   C. Guaraldo, T. Hashimoto, R. S. Hayano, T. Hiraiwa, M. Iio,
   M. Iliescu, K. Inoue, Y. Ishiguro, T. Ishikawa, S. Ishimoto,
   T. Ishiwatari, K. Itahashi, M. Iwai, M. Iwasaki, Y. Kato,
   S. Kawasaki, P. Kienle, H. Kou, J. Marton, Y. Matsuda,
   Y. Mizoi, O. Morra, T. Nagae, H. Noumi, H. Ohnishi,
   S. Okada, H. Outa, K. Piscicchia, M. Poli Lener,
   A. Romero Vidal, Y. Sada, A. Sakaguchi, F. Sakuma, M. Sato,
   A. Scordo, M. Sekimoto, H. Shi, D. Sirghi, F. Sirghi, K. Suzuki,
   S. Suzuki, T. Suzuki, K. Tanida, H. Tatsuno, M. Tokuda,
   D. Tomono, A. Toyoda, K. Tsukada, O. Vazquez Doce,
   E. Widmann, B. K. Wuenschek, T. Yamaga, T. Yamazaki,
   H. Yim, Q. Zhang and J. Zmeskal   107

Studying the $e^+e^- \rightarrow (D^*\bar{D}^*)^\pm\pi^\mp$ reaction and the claim for
the $Z_c(4025)$ resonance
   A. Martínez Torres, K. P. Khemchandani, F. S. Navarra,
   M. Nielsen and E. Oset   111

Life on earth – An accident? Chiral symmetry and the anthropic
principle
   U.-G. Meißner   116

Decays of doubly charmed meson molecules
   R. Molina, A. Hosaka and H. Nagahiro   124

Chiral symmetry breaking and restoration with mixing between
quarkonium and tetraquark
   T. K. Mukherjee and M. Huang   128

Double-Λ hypernuclei at J-PARC − E07 experiment
K. Nakazawa and J. Yoshida   132

Regge trajectory of the $f_0(500)$ resonance from a dispersive connection to its pole
J. Nebreda, J. T. Londergan, J. R. Pelaez and A. P. Szczepaniak   137

Recent developments on LQCD studies of nuclear force
H. Nemura   141

Spectroscopy of heavy quark hadrons
M. Oka   149

Recent developments on hadron interaction and dynamically generated resonances
E. Oset, M. Albaladejo, J.-J. Xie and A. Ramos   157

Power counting in nuclear effective field theory
M. Pavon Valderrama   165

Present status of light scalars
J. R. Peláez   169

Hyperon-nucleon interaction and baryonic contact terms in SU(3) chiral effective field theory
S. Petschauer   177

Octet baryon masses and sigma terms in covariant baryon chiral perturbation theory
X.-L. Ren, L.-S. Geng and J. Meng   181

Roy–Steiner equations for πN scattering
J. Ruiz de Elvira, C. Ditsche, M. Hoferichter, B. Kubis and U.-G. Meißner   186

Experimental results on $Z_c(3900)$
C. P. Shen   190

Dynamically generated resonances from the vector octet-baryon
octet interaction with strangeness zero
    B.-X. Sun, J. Wang and X.-F. Lu   194

Massive hybrid stars with strangeness
    T. Takatsuka, T. Hatsuda and K. Masuda   198

Shape-phase transitions in very neutron-rich nuclei from
$_{40}$Zr to $_{46}$Pd
    H. Watanabe   202

New hidden beauty molecules predicted by the local hidden
gauge approach and heavy quark spin symmetry
    C. W. Xiao, A. Ozpineci and E. Oset   206

Understanding the negative parity Λ resonances from the
$K^-p \to \Lambda\eta$ reaction and their strong decays
    L.-Y. Xiao and X.-H. Zhong   210

Exotic dibaryons with a heavy antiquark
    Y. Yamaguchi, A. Hosaka and S. Yasui   214

Recent progress towards a chiral effective field theory for the
NN system
    C. J. Yang and B. Long   218

Progress in resolving charge symmetry violation in nucleon
structure
    R. D. Young, P. E. Shanahan and A. W. Thomas   222

Chirality in atomic nuclei: 2013
    S. Q. Zhang, Q. B. Chen and J. Meng   230

Preface

The Seventh International Symposium on Chiral Symmetry in Hadrons and Nuclei (Chiral 2013) was held at Beihang University (Beijing, China) on October 27–30 2013. This follows a series of symposia that was first held at Yukawa Institute for Theoretical Physics, Kyoto, Japan, in 2000. The subsequent 5 symposia were then held at Orsay (2001), Kyoto (2002), Tokyo (2005), Osaka (2007), and Valencia (2010).

As a regular event now in the hadron and nuclear physics community, the symposium has been an important platform for revealing the recent progresses in hadron and nuclear physics related to chiral symmetry In particular, it brings together expertise from both fields which can inspire developments of new theoretical tools in order to tackle the fundamental aspects of chiral symmetry. We have seen such efforts made by the speakers in their presentations at the symposium and we have enjoyed greatly the active atmosphere created by the participants. About 100 participants from 15 countries in 5 continents attended the four-day symposium, and there were 19 plenary talks and 46 parallel talks presented at the meeting which covered broad topics in hadron and nuclear physics:

- Chiral and heavy-quark spin symmetry
- Chiral dynamics of few-body hadron systems
- Chiral symmetry and hadrons in a nuclear medium
- Chiral dynamics in nucleon-nucleon interaction and atomic nuclei
- Chiral symmetry in rotating nuclei
- Hadron structure and interactions
- Exotic hadrons, heavy flavor hadrons and nuclei
- Mesonic atoms and nuclei

All the talks (in form of PDF or PPT) have been published on the website of the symposium: http://indico.ihep.ac.cn/conferenceDisplay.py?confId=2948

We appreciate the IAC members for their advice on the scientific program and thank all the participants and supporting staffs for their great contributions that have made this symposium a success. We would also like to acknowledge the financial support from the following institutions, without which this event would have not been possible to happen:

- Beihang University,
- China Center of Advanced Science and Technology
- Hunan Normal University
- Institute of Theoretical Physics (CAS)
- National Natural Science Foundation of China,
- Theoretical Physics Center for Science Facilities (CAS)
- Tsinghua University

Editors:

Li-Sheng Geng
School of Physics and Nuclear Energy Engineering, Beihang University, Beijing 100191, China
E-mail: lisheng.geng@buaa.edu.cn

Jie Meng
School of Physics, Peking University, Beijing 100871, China
E-mail: mengj@pku.edu.cn

Qiang Zhao
Institute of High Energy Physics, Chinese Academy of Sciences, Beijing 100049, China
E-mail: zhaoq@ihep.ac.cn

Bing-Song Zou
Institute of Theoretical Physics, Chinese Academy of Sciences, Beijing 100190, China
E-mail: zoubs@itp.ac.cn

The Seventh International Symposium on Chiral Symmetry in Hadrons and Nuclei
October 27th —30th, 2013, Beijing, China

Studies of chirality in the MASS 80, 100 and 190 regions

R. A. Bark*, E. O. Lieder, R. M. Lieder, E. A. Lawrie, J. J. Lawrie, S. P. Bvumbi,
N. Y. Kheswa, S. S. Ntshangase, T. E. Madiba, P. L. Masiteng, S. M. Mullins,
S. Murray, P. Papka, O. Shirinda

iThemba LABS, P.O. Box 722, Somerset West 7129, South Africa
* E-mail: bark@tlabs.ac.za

Q. B. Chen, S. Q. Zhang, Z. H. Zhang, P. W. Zhao, C. Xu

State Key Laboratory of Nuclear Physics and Technology, School of Physics,
Peking University, Beijing 100871, China

J. Meng

State Key Laboratory of Nuclear Physics and Technology, School of Physics,
Peking University, Beijing 100871, China
School of Physics and Nuclear Energy Engineering, Beihang University,
Beijing 100191, China
Department of Physics, University of Stellenbosch, Stellenbosch, South Africa

D. G. Roux

Department of Physics and Electronics, Rhodes University, Grahamstown 6410,
South Africa

Z. P. Li

School of Physical Science and Technology, Southwest University, Chongqing 400715,
China

J. Peng

Department of Physics, Beijing Normal University, Beijing 100875, China

B. Qi

Shandong Provincial Key Laboratory of Optical Astronomy and Solar-Terrestrial
Environment, School of Space Science and Physics, Shandong University,
Weihai 264209, China

S. Y. Wang

Shandong Provincial Key Laboratory of Optical Astronomy and Solar-Terrestrial Environment, School of Space Science and Physics, Shandong University, Weihai 264209, China

Z. G. Xiao

Department of Physics, Tsinghua University, Beijing 100084, China Collaborative Innovation Center of Quantum Matter, Beijing, China

A brief survey of results of studies of nuclear chirality in the mass 80 and 190 region at iThemba LABS is given, before looking at the case of ^{106}Ag in detail. Here, the crossing of a pair of candidate chiral bands is re-interpreted as the crossing of a prolate band by an aligned four-quasiparticle band.

Keywords: ^{96}Zr(^{14}N,4n); ^{106}Ag deduced levels; J; π; chiral bands; transition rates; configurations; alignments.

1. Introduction

It was back in 1997 that Frauendorf and Meng[1] suggested the possibility of observing chiral symmetry in the angular momentum space of atomic nuclei. If a system could be found where the angular momentum of a rotating triaxial nucleus could be coupled to orthogonal proton and neutron angular momenta, the resulting total angular momentum would be aplanar (lying outside of the plane defined by any two of the above angular momentum vectors) and it would be possible to define left and right-handed systems. The left and right-handed systems would, in the laboratory frame, be characterized by rotational bands of identical properties such as energies and transition strengths. Their paper sparked numerous experimental investigations to find examples of nuclear chirality and to test the chiral hypothesis. Today there are many known examples, with those in ^{126}Cs being among the best.[2] Nevertheless, many open questions remain.

To what extent is nuclear chirality universal? Some of the first examples of nuclear chirality were found in the odd-odd nuclei of the mass 130 region,[3] where $h_{11/2}$ protons couple to $h_{11/2}$ neutrons. But to what extent does chirality exist in other mass regions, with other configurations, and with different quasiparticle multiplicities?

To what degree can degeneracy actually be reached in atomic nuclei? This question requires complete spectroscopy to be answered - not only must the energies of band members be measured, but also the transition rates between the levels of the bands. In odd-odd nuclei, a complicating factor is that these quantities are also expected to be influenced by residual proton-neutron interactions.

At iThemba LABS, a research programme has been in place for a number of years to examine these questions. We present results, also obtained in collaboration with colleagues from China and Hungary, from experiments using the AFRODITE array[4] of up to 9 Compton-suppressed "clover" γ-ray detectors, and the DIAMANT CsI array of charged-particle detectors.

2. New Regions: 80 and 190

New regions of chirality have been identified using AFRODITE. In Wang et al.,[5] a pair of bands, assigned to the $\pi g_{9/2} \otimes \nu g_{9/2}$ configuration, were identified as the first examples of chirality in the mass 80 region. This programme has now been extended by a recent experiment to the study ^{78}Br, which is reported in detail in another talk at this meeting.

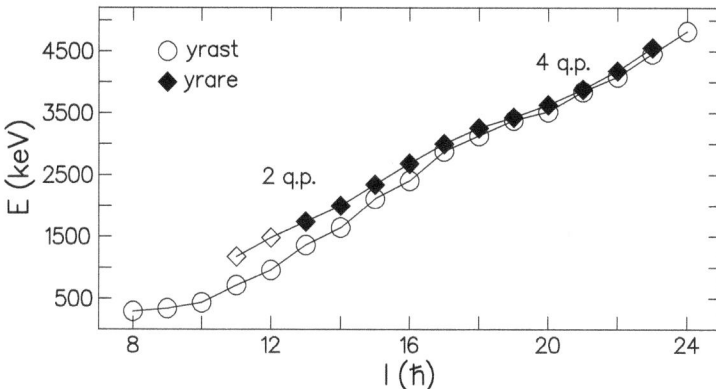

Fig. 1.   Comparison of the energies of the yrast and side bands in ^{194}Tl.Degeneracy is reached above spin 16.

While the first chiral candidate pair in the mass 190 region was identified by Balabanski et al.[6] in ^{188}Ir, our work in this region has concentrated on studying the isotopes of Thalium. This provided evidence of the effect of residual proton-neutron interactions on the degeneracy of the partner bands of the $\pi h_{9/2} \otimes \nu i_{13/2}$ configuration in ^{198}Tl.[7] Perhaps the most spectacular example of near-degeneracy of chiral partners was found in ^{194}Tl.[8] Two bands, assigned to the $\pi h_{9/2} \otimes \nu i_{13/2}^3$ four quasiparticle configuration were identified. The energies of the levels of the bands are shown in Figure 1. Two, 2-quasiparticle $\pi h_{9/2} \otimes \nu i_{13/2}$ bands are crossed at spin 16 by two 4-quasiparticle $\pi h_{9/2} \otimes \nu i_{13/2}^3$ configurations. Above this spin, the relative excitation energy of the partners remains less than 110 keV. Other quantities, such as B(M1)/B(E2) transition rate ratios, are also very close in value, in what is perhaps the best known example of chiral degeneracy.

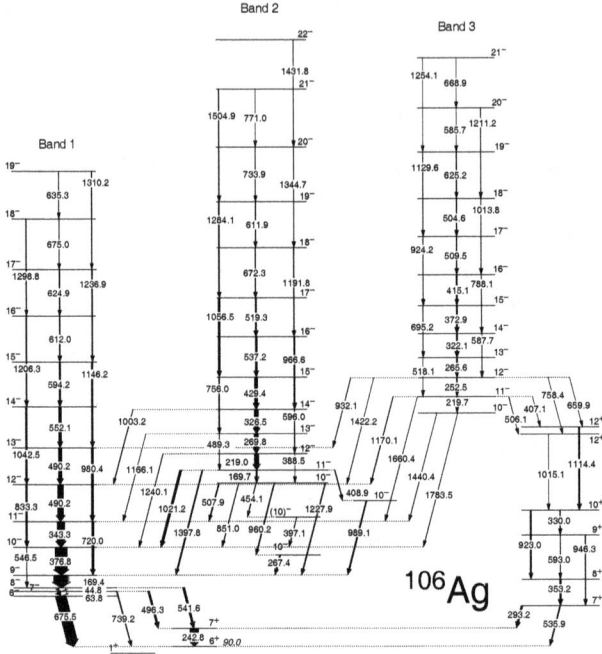

Fig. 2.   The level scheme of ^{106}Ag.

3. Complete Spectroscopy: Ag-106

The nucleus ^{106}Ag is an interesting case in which Joshi et al.[9] first identified a pair of negative-parity bands, assigned to the $\pi g_{9/2} \otimes \nu h_{11/2}$ configuration, that actually become degenerate and cross each other near spin 14. Interestingly, Joshi et al. hypothesized that the lowest energy band correspond to a triaxial shape, while the partner band have a prolate shape, induced by a chiral vibration. To test this hypothesis more closely, a detailed investigation was performed at iThemba LABS. A number of measurements were performed using the ^{96}Zr(^{14}N,4n) reaction. Thin target (0.7 and 1.6 mg/cm^2) γ-ray coincidence experiments allowed the level scheme to be revised and extended, and allowed polarization anisotropies and angular correlations of γ-ray transitions to be measured. In a third experiment, a thick 14 mg/cm^2 target was used, which allowed the Doppler Shift Attenuation Method (DSAM) to be used to determine level lifetimes.[10–12]

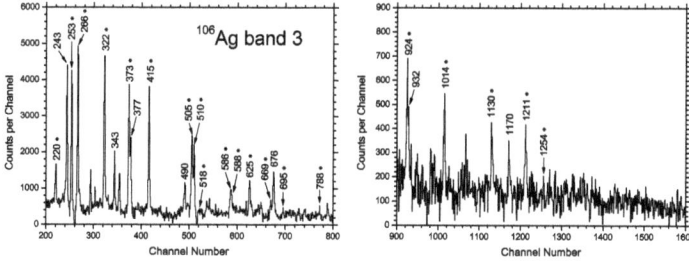

Fig. 3. Coincidence spectra obtained by gating on the 253, 266, 322 and 373 keV lines of band 3. Lines labelled with an asterix denote transitions placed in band 3.

The level scheme obtained from this investigation is shown in Figure 2; compared to prior works,[9,13,14] it has been extended to higher spin and some levels have been re-arranged. Bands 1 and 2 are the two bands identified by Joshi et al., but band 3 is also of negative parity. Spectra obtained by setting coincidence windows on transitions at the bottom of band 3, using the thin target (0.7 mg/cm^2) data are shown in Figure 3. They show for example new transitions in the band, decaying from levels placed above the spin 15$^-$ state.

One way to test the identity of bands 1 to 3 is to perform a quasi-particle alignment analysis. The component of angular momentum aligned with the axis of rotation is given in the Cranked Shell Model,[15] by $I_x = \sqrt{I(I+1) - K^2}$, from which a core contribution is subtracted to obtain the quasiparticle alignment i_x. The γ-ray transition defines a rotational frequency, by $\omega = E_\gamma/2$. The alignment i_x is additive: by empirically determining the contribution of the constituent quasiparticles from neighbouring odd-mass nuclei, one can predict the expected alignments for various configurations in ^{106}Ag. This has been done in Figure 4 for bands 1, 2, and 3. One observes that the alignment of band 1 is in good agreement with its original assignment to a $\pi g_{9/2} \otimes \nu h_{11/2}$ configuration. However, the alignment of band 2 is inconsistent with this configuration. Rather, both bands 2 and 3 are consistent with the four-quasiparticle configuration

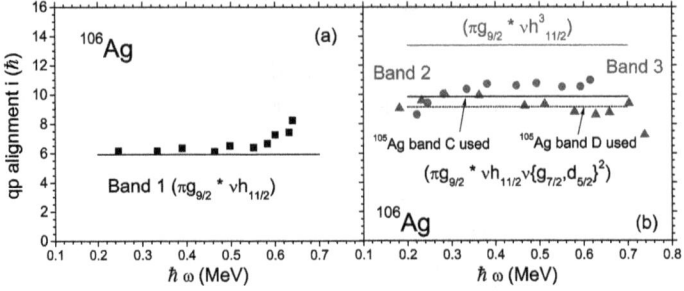

Fig. 4. Symbols: experimental quasiparticle alignment as a function of rotational freqency for bands 1 to 3. Lines: alignments deduced empirically from neighbouring nuclei.

$\pi g_{9/2} \otimes \nu h_{11/2} \nu \left\{ g_{7/2}, d_{5/2} \right\}^2$, where $\left\{ g_{7/2}, d_{5/2} \right\}$ indicates a mixture of $g_{7/2}$ and $d_{5/2}$ orbitals. This would imply then, that bands 1 and 2 of ^{106}Ag are not chiral partners at all, having different numbers of quasiparticles.

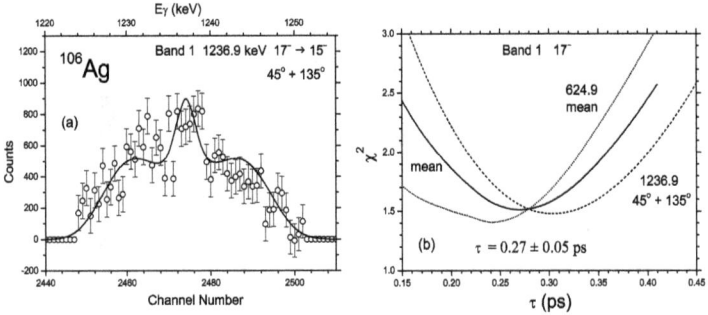

Fig. 5. (a) Solid is a DSAM analysis fit to the summed lineshape of the 1236.9 keV transition. (b) Chi-squared analysis determining the meanlife of the 17^- level of band 1.

An example of the DSAM analysis is shown in Figure 5 for the 17^- level of band 1. In this measurement, four clover detectors were placed at $45°$ and four at $135°$. In panel (a), the lineshape of the 1236.9 keV transition, which depopulates the 17^- level, is shown, summing contributions from both angles. As the recoiling nucleus slows down in the target, γ-rays emitted from moving nuclei are shifted to higher(lower) energies if observed in the $45°(135°)$ detectors, giving rise to a broad distribution of energies around the central (unshifted) peak. A knowledge of the stopping process together with the lineshape allows the lifetime of the level to be fitted. The resulting chi-squared distributions are shown for the 1236.9 and 624.9 keV transitions from the 17^- level in the lower panel. The combination yields a level lifetime of 0.27 ±0.05 ps.

The lifetimes allow reduced transition probabilities to be extracted and compared with the results of particle-rotor model calculations in Figure 6. Overall a good agreement is obtained for the newly proposed configurations.

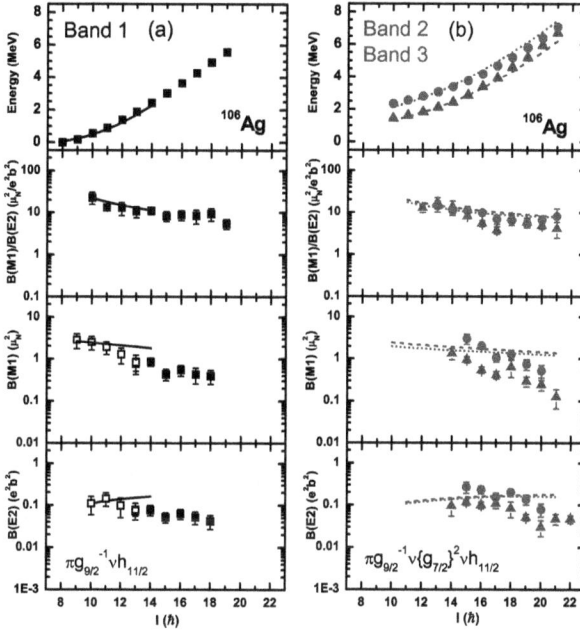

Fig. 6. Measured properties of bands 1, 2 and 3 compared with results of particle-rotor model calculations.

Acknowledgments

The authors thank the iThemba LABS technical staff and accelerator group for their support. R.M.L. wants to thank iThemba LABS and the Peking University for financial support during short term stays at iThemba LABS and the State Key Laboratory of Nuclear Physics and Technology, respectively. This work was supported by the National Research Foundation of South Africa, the Major State 973 Program of China (Grant No. 2013CB834400), the Natural Science Foundation of China (Grants No. 11175002, No. 11335002 and No. 11375015), the Research Fund for the

Doctoral Program of Higher Education (Grant No. 20110001110087), and the China Postdoctoral Science Foundation (Grants No. 2012M520101 and No. 2013M540011).

References

1. S. Frauendorf and J. Meng, *Nucl. Phys. A* **617**, 131 (1997).
2. E. Grodner *et al.*, *Eur. J. Phys. A* **42**, 70 (2009).
3. K. Starosta, T. Koike, C.J. Chiara, D.B. Fossan, D.R. LaFosse, A.A. Hecht, C.W. Beausang, M.A. Caprio, J.R.Cooper, R. Krücken, J.R. Novak, N.V. Zamfir, K.E. Zyromski, D.J. Hartley, D.L. Balabanski, Jing-ye Zhang, S. Frauendorf, and V.I. Dimitrov, Phys. Rev. Lett. **86**, 971 (2001).
4. R.T. Newman *et al.*, *Balkan Phys. Lett.*, **Special Issue**, 182 (1998)
5. S.Y. Wang, B. Qi, L. Liu, S.Q. Zhang, H. Hua, X.Q. Li, Y.Y. Chen, L.H. Zhu, J. Meng, S.M. Wyngaardt, P. Papka, T.T. Ibrahim, R.A. Bark, P. Datta, E.A. Lawrie, J.J. Lawrie, S.N.T. Majola, P.L. Masiteng, S.M. Mullins, J. Gal, G. Kalinka, J. Molnar, B.M. Nyako, J. Timar, K. Juhasz, and R. Schwengner, *Phys. Lett. B* **703**, 40 (2011).
6. D.L. Balabanski *et al.*, *Phys. Rev. C* **70**, 044305 (2004)
7. E.A. Lawrie, P.A. Vymers, J.J. Lawrie, Ch. Vieu, R.A. Bark, R. Lindsay, G.K. Mabala, S.M. Maliage, P.L. Masiteng, S.M. Mullins, S.H.T. Murray, I. Ragnarsson, T.M. Ramashidzha, C. Schück, J.F. Sharpey-Schafer, and O. Shirinda, *Phys. Rev. C* **78**, 021305(R) (2008).
8. P.L. Masiteng, E.A. Lawrie, T.M. Ramashidzha, R.A. Bark, B.G. Carlsson,J.J. Lawrie, R. Lindsay, F. Komati, J. Kau, P. Maine, S.M. Maliage, I. Matamba, S.M. Mullins, S.H.T. Murray, K.P. Mutshena, A.A. Pasternak, I. Ragnarsson, D.G. Roux, J.F. Sharpey-Schafer, O. Shirinda, and P.A. Vymers, *Phys. Lett. B* **719**, 83 (2013).
9. P. Joshi, M.P. Carpenter, D.B. Fossan, T. Koike, E.S. Paul, G. Rainovski, K. Starosta, C. Vaman, and R. Wadsworth, *Phys. Rev. Lett.* **98**, 102501 (2007).
10. E. Grodner, A.A. Pasternak, Ch. Droste, T. Morek, J. Srebrny, J. Kownacki, W. Płóciennik, A.A. Wasilewski, M. Kowalczyk, M. Kisieliński, R. Kaczarowski, E. Ruchowska, A. Kordyasz, and M. Wolińska, Eur. Phys. J. A **27**, 325 (2006).
11. E.O. Lieder, A.A. Pasternak, R.M. Lieder, A.D. Efimov, V.M. Mikhajlov, B.G. Carlsson, I. Ragnarsson, W. Gast, Ts. Venkova, T. Morek, S. Chmel, G. de Angelis, D.R. Napoli, A. Gadea, D. Bazzacco, R. Menegazzo, S. Lunardi, W. Urban, Ch. Droste, T. Rząca-Urban, G. Duchêne, and A. Dewald, *Eur. Phys. J. A* **35**, 135 (2008).
12. R.M. Lieder *et al.*, *Eur. Phys. J. A* **21**, 37 (2004).
13. A.I. Levon, J. de Boer, A.A. Pasternak, and D.A. Volkov, *Z. Phys. A* **343**, 131 (1992).
14. D. Jerrestam, W. Klamra, J. Gizon, F. Lidén, L. Hildingsson, J. Kownacki, Th. Lindblad, and J. Nyberg, *Nucl. Phys. A* **577**, 786 (1994).
15. R. Bengtsson and S. Frauendorf, *Nucl. Phys. A* **327**, 139 (1979).

$\rho K \bar{K}$ system within the framework of the fixed center approximation to Faddeev equations

M. Bayar

(1)-Departamento de Física Teórica and IFIC, Centro Mixto Universidad de Valencia-CSIC,
Institutos de Investigación de Paterna, Apartado 22085, 46071 Valencia, Spain
(2) Department of Physics, Kocaeli University, 41380, Izmit, Turkey
E-mail: melahat.bayar@kocaeli.edu.tr
http://www.kocaeli.edu.tr/int/

W. Liang

(1) and (3) Department of Physics, Guangxi Normal University,
Guilin, 541004, P. R. China
E-mail: liangwh@mailbox.gxnu.edu.cn

T. Uchino and C. W. Xiao

(1) E-mail: toshitaka.uchino@gmail.com, xiaochw@ific.uv.es

We perform a calculation for three body $\rho K \bar{K}$ scattering amplitude by using the fixed-center approximation to Faddeev Equations, taking the interaction between ρ and $\bar{K}$, ρ and K from the chiral unitary approach. We find peak in the modulus squared of the three-body scattering amplitude, indicating the existence of resonance which can be associated to $\rho(1700)$.

Keywords: Dynamically generated resonances; multi-hadron systems.

1. Introduction

At low energies, the dynamics of light hadrons can be described in terms of chiral symmetry.[1,2] By using the leading order of the chiral Lagrangians as input, a powerful tool, the chiral unitary approach, implementing unitarity in coupled channels, has been developed and it has provided great success describing many resonances for meson-meson or meson-baryon systems.[3,4]

In order to explore multi-hadron systems, the application of the fixed center approximation to Faddeev equations has been implemented.[5,6] Under the condition where the cluster structure in three-body systems is not varied so much against the collision of the other particle, that approximation will work fine, as discussed in Refs. 7, 8.

2. ρK Unitarized Amplitude

In order to solve the Faddeev equation within the fixed center approximation, one needs the two-body unitarized amplitude. In the present case of the $\rho K \bar{K}$ system, the ρK ($\rho \bar{K}$) unitarized amplitude is necessary. In the previous work[9,10] the vector-pseudoscalar interaction in the sector, strangeness $S = 1$ and isospin $I = 1/2$ was studied within the framework of the chiral unitary approach and that interaction was shown to generate two resonance poles corresponding to the $K_1(1270)$ resonance. The detailed calculation has been done.[11]

3. The $\rho K \bar{K}$ Three-body Scattering

Once the unitarized ρK amplitude is obtained, let us go to the $\rho K \bar{K}$ three-body system. As mentioned above, we study this system by solving the Faddeev equation within the fixed center approximation. For the detailed formalism one can look at the original paper.[11]

Under the fixed center approximation, it is assumed that two particles 1 and 2 cluster together and the structure of the cluster is kept against the collision of the extra particle 3. This idea leads the three-body scattering amplitude T which can read a summation of the two following partition functions T_1 and T_2

$$
\begin{aligned}
T_1 &= t_1 + t_1 G_0 T_2, \\
T_2 &= t_2 + t_2 G_0 T_1, \\
T &= T_1 + T_2,
\end{aligned}
\tag{1}
$$

where the subscripts $i = 1, 2$ of T_i and t_i represent the component particle i in the cluster. The interaction kernel t_i between the particle i and 3 stands for the leading order contribution in the partition function T_i. From the double scattering, the propagation of the particle 3 appears and is described as a function G_0. The G_0 function of eq. (1) reads as a function of the energy $\sqrt{s}$

$$
G_0(\sqrt{s}) = \frac{1}{2M_{f_0}} \int \frac{d^3 q}{(2\pi)^3} F_{f_0}(q) \frac{1}{q^{02}(s) - \vec{q}^2 - m_\rho^2 + i\epsilon},
\tag{2}
$$

where M_{f_0} is the mass of the $f_0(980)$ resonance and the form factor

$$F_{f_0}(q) = \frac{1}{\mathcal{N}} \int_{\substack{p<k_{max} \\ |\vec{p}-\vec{q}|<k_{max}}} d^3p \left(\frac{1}{2\omega_K(\vec{p})}\right)^2 \frac{1}{M_{f_0} - 2\omega_K(\vec{p})}$$

$$\times \left(\frac{1}{2\omega_K(\vec{p}-\vec{q})}\right)^2 \frac{1}{M_{f_0} - 2\omega_K(\vec{p}-\vec{q})}, \tag{3}$$

where the normalization $\mathcal{N}$ is given by

$$\mathcal{N} = \int_{p<k_{max}} d^3p \left[\left(\frac{1}{2\omega_K(\vec{p})}\right)^2 \frac{1}{M_{f_0} - 2\omega_K(\vec{p})}\right]^2. \tag{4}$$

4. Results

The $\rho K\bar{K}$ three-body amplitude within the fixed center approximation is calculated. To see the effect of the multiple scattering, the full amplitude is compared with the single scattering amplitude. We also consider the width of the ρ, $\Gamma_\rho \sim 150$ MeV, in the G_0 function. By using the following replacement for the ρ propagator in eq. (2)

$$\frac{1}{q^{02} - \vec{q}^2 - m_\rho^2 + i\epsilon} \rightarrow \frac{1}{q^{02} - \vec{q}^2 - m_\rho^2 + im_\rho\Gamma_\rho}, \tag{5}$$

we include the ρ width into the G_0 function. In fig. 1 (left-hand side), the $\rho^+[K\bar{K}]_{I=0}$ amplitude is shown and there one can see that a peak appears in each case. In the single scattering amplitude, the peak exists above the threshold of ρ and $f_0(980)$. Through the multiple scattering, the peak position shifts lower while the width is getting smaller because the $\rho f_0(980)$ channel becomes closed. With the ρ width effect, that peak is getting much wider while the shift of the peak position is not so large. The masses or a peak position in the amplitude (full widths) at half maximum of the dynamically generated state are 1777.9 MeV (144.4 MeV), 1734.8 MeV (63.7 MeV) and 1748.0 MeV (160.8 MeV) for the single scattering, full scattering, full scattering with the ρ width effect, respectively. The experimental data in PDG[13] is 1720 ± 20 MeV (250 ± 100 MeV). Compared with the experimental data, it is found to be important to take into account the ρ width.

As shown above, the present study seems to work for generating the resonance which might correspond to $\rho(1700)$. In addition, we consider the width of the $f_0(980)$ too. The $K\bar{K}$ component in the $f_0(980)$ resonance is found to be dominant while the decay width into the $\pi\pi$ channel is

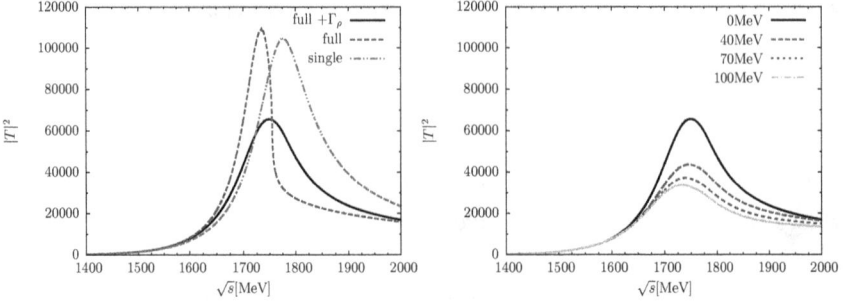

Fig. 1. The $\rho K \bar{K}$ amplitude where "single", "full", "full + Γ_ρ" denote the amplitude of the single scattering, full scattering, full scattering with the ρ width effect respectively (left-hand side) and the $\rho K \bar{K}$ amplitude with the ρ and $f_0(980)$ width effect (right-hand side).

not so small. Therefore we consider the inclusion of the width into our framework. In order to keep the fixed center approximation, here we give a naive prescription that the eigenvalue of the $K \bar{K}$ system is now a complex value. Namely the mass of the cluster M_{f_0} in eqs. (3) and (4) is replaced by $M_{f_0} - i\Gamma_{f_0}/2$. Then the form factor becomes a complex function which might represent the effect of the wave function of the unstable cluster. The amplitude with the $f_0(980)$ and ρ width effect is shown in fig. 1 (right-hand side). The masses (widths) of the dynamically generated state are 1748.0 MeV (160.8 MeV), 1743.6 MeV (216.4 MeV), 1739.2 MeV (227.2 MeV) and 1734.8 MeV (224.6 MeV) for $\Gamma_{f_0} = 0, 40, 70, 100$ MeV, respectively. Taking into account the ambiguity of the $f_0(980)$ width, as a value of Γ_{f_0}, a maximum and minimum value of the experimental data and their average are taken. It is shown that the inclusion of the $f_0(980)$ width induces a suppression of the magnitude of the peak and the peak becomes broader as the width of the $f_0(980)$ increases. Furthermore it is also a remarkable feature that the peak position is not so affected by this prescription.

Acknowledgments

This work is partly supported by the Spanish Ministerio de Economia y Competitividad and European FEDER funds under the Contract No. FIS2011-28853-C02-01 and the Generalitat Valenciana in the program Prometeo, 2009/090. We acknowledge the support of the European Community-Research Infrastructure Integrating Activity Study of Strongly Interacting Matter(acronym Hadron Physics 3, Grant No. 283286) under the Seventh Framework Programme of the European Union. This work is also partly supported by the National Natural Science Foundation of China under Grant No. 11165005 and by scientific research fund (201203YB017) of Education Department of Guangxi.

References

1. J. Gasser and H. Leutwyler, Nucl. Phys. B **250**, 465 (1985).
2. U. G. Meissner, Rept. Prog. Phys. **56**, 903 (1993)
3. J. A. Oller and E. Oset, Nucl. Phys. A **620**, 438 (1997) [Erratum-ibid. A **652**, 407 (1999)]
4. E. Oset and A. Ramos, Nucl. Phys. A **635**, 99 (1998)
5. L. D. Faddeev, Sov. Phys. JETP **12**, 1014 (1961) [Zh. Eksp. Teor. Fiz. **39**, 1459 (1960)].
6. A. Gal, Int. J. Mod. Phys. A **22**, 226 (2007)
7. A. Martinez Torres, E. J. Garzon, E. Oset and L. R. Dai, Phys. Rev. D **83**, 116002 (2011)
8. M. Bayar, J. Yamagata-Sekihara and E. Oset, Phys. Rev. C **84**, 015209 (2011)
9. L. Roca, E. Oset and J. Singh, Phys. Rev. D **72**, 014002 (2005)
10. L. S. Geng, E. Oset, L. Roca and J. A. Oller, Phys. Rev. D **75**, 014017 (2007)
11. M. Bayar, W. H. Liang, T. Uchino and C. W. Xiao, arXiv:1312.2869 [hep-ph].
12. W. Liang, C. W. Xiao and E. Oset, arXiv:1309.7310 [hep-ph].
13. J. Beringer *et al.* [Particle Data Group Collaboration], Phys. Rev. D **86**, 010001 (2012).

Symmetry breaking in the parton distribution functions of the nucleon

F.-G. Cao

Institute of Fundamental Sciences, Massey University,
Private Bag 11-222, Palmerston North, New Zealand
E-mail: f.g.cao@massey.ac.nz

Parton models predict several symmetries for the parton distribution functions of the nucleon, including flavour symmetry, charge symmetry, and quark-antiquark symmetry. We report calculations using the meson cloud model for the breaking of these symmetries.

Keywords: Symmetry breaking; parton distribution function; meson cloud model.

1. Introduction

Parton distribution functions (PDFs) are related to the distributions defined in the parton models at leading order with logarithmic scaling violations. According to the parton models there are various symmetries for the PDFs of the nucleon, including SU(2) flavour symmetry ($\bar{d} = \bar{u}$), charge symmetry (the invariance of a system under the interchange of up and down quarks, e.g. $\bar{u}^p = \bar{d}^n$, and $\bar{d}^p = \bar{u}^n$) and quark-antiquark symmetry ($s = \bar{s}$).

The breaking of flavour symmetry in the unpolarized proton sea, $\bar{d} > \bar{u}$, has been well established experimentally[1] and this asymmetry can be naturally explained in the meson cloud model.[2] We report investigations on the breaking of flavour symmetry,[3] charge symmetry,[4] and quark-antiquark symmetry[5,6] in the nucleon's PDFs using the meson cloud model.

2. Meson Cloud Model

In the meson cloud model the nucleon can be viewed as a baryon 'core' surrounded by a mesonic cloud. The lifetime of the virtual baryon-meson components is generally much longer than the interaction time in the deep inelastic process, thus the quarks and antiquarks in the baryon and meson contribute to the PDFs of the nucleon. The observed PDFs are the sum of contributions from the 'bare' nucleon and the virtual baryons and mesons. The contribution from the virtual baryons can be expressed as a convolution of fluctuation function $f_{BM/N}(y)$, which describes the probability for a nucleon fluctuating into a virtual baryon (B) with longitudinal momentum fraction y and a virtual meson (M) with longitudinal momentum fraction $1 - y$, with the valence PDF of the baryon,

$$\delta q(x, Q^2) = \int_x^1 \frac{dy}{y} f_{BM/N}(y) q_B(\frac{x}{y}, Q^2).$$ (1)

The contribution from the virtual mesons can be expressed in a similar way.

Considering these contributions from the virtual baryons and mesons leads to the breaking of various symmetries predicted by the parton models. The flavour asymmetry in the proton $\bar{d} > \bar{u}$ emerges since the probability of finding the $|n\pi^+\rangle$ component in the proton is larger than that of the $|\Delta^{++}\pi^-\rangle$ component.

3. Symmetry Breaking

3.1. *Flavour symmetry breaking in the polarized nucleon sea*

The theoretical calculations for the difference $x[\Delta\bar{u}(x) - \Delta\bar{d}(x)]$ and the experimental results from the HERMES Collaboration[7] are shown in Fig. 1. The theoretical calculations are consistent with the data, although large uncertainties exist in the data. The SU(2) flavour symmetry breaking in the polarized nucleon sea is much smaller than that in the unpolarized sea.

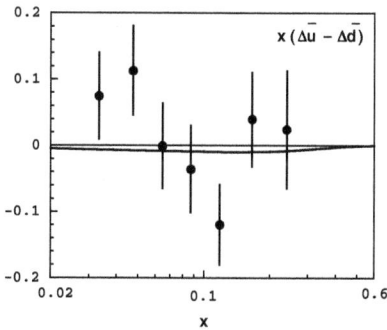

Fig. 1. Flavour symmetry breaking in the polarized light antiquark sea. The solid curve is the theoretical calculation. The HERMES data are taken from.[7]

3.2. *Charge symmetry breaking*

The breaking of charge symmetry in the PDFs of the nucleon can be 'measured' via the quantities,

$$\delta d_v(x) = d_v^p(x) - u_v^n(x), \; \delta u_v(x) = u_v^p(x) - d_v^n(x),$$
$$\delta \bar{d}(x) = \bar{d}^p(x) - \bar{u}^n(x), \; \delta \bar{u}(x) = \bar{u}^p(x) - \bar{d}^n(x). \tag{2}$$

The results for the charge symmetry breaking in the valence quarks ($x\delta d_v$ and $x\delta u_v$) are shown in Fig. 2. The distributions $x\delta d_v$ and $x\delta u_v$ have similar shape and both are negative. We did not find any significantly large charge symmetry breaking in the sea quark distributions of the nucleon.

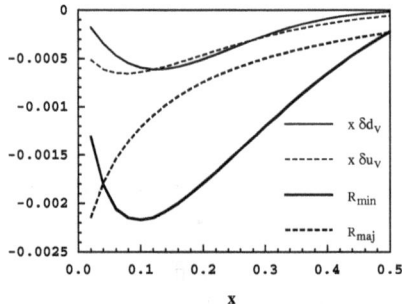

Fig. 2. The charge symmetry breaking in the valence quarks. $R_{\min} = \delta d_v/d_v^p$ and $R_{\text{maj}} = \delta u_v/u_v^p$. $Q^2 = 4$ GeV^2.

3.3. *Strange-antistrange symmetry breaking*

There are two mechanisms for the generation of an asymmetry between the strange and antistrange quark distributions in the nucleon: nonperturbative contributions originating from nucleons fluctuating into virtual baryon-meson pairs such as ΛK and ΣK, and perturbative contributions arising from gluons splitting into strange and anti-strange quark pairs.

The nonperturbative contributions can be calculated using the meson cloud model while the perturbative contributions are the result of considering QCD evolution at next-to-next-to-leading order. The results are presented in Fig. 3. The perturbative and nonperturbative contributions are dominant in the small- and large-x regions, respectively.

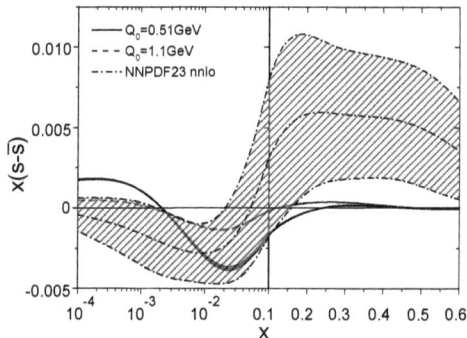

Fig. 3.   The total asymmetry $x(s-\bar{s})$ at $Q^2 = 16$ GeV^2. Only the K-meson contributions are considered in the nonperturbative calculations. The solid and dashed curves are the results obtained with $Q_0 = 0.51$ GeV and $Q_0 = 1.1$ GeV in the perturbative calculations.

4. Summary

Using the meson cloud model we calculated the possible breaking of symmetries predicted by the parton models. We found that the SU(2) flavour symmetry breaking in the polarized nucleon sea is much smaller than that in the unpolarized sea, charge symmetry breaking is smaller than 1%, and the strange-antistrange asymmetry may have more than one node.

References

1. A. Baldit et al., NA51 Collaboration, Phys. Lett. B **332**, 244 (1994); E. A. Hawker et al., E866/NuSea Collaboration, Phys. Rev. Lett. **80**, 3715 (1998); K. Ackerstaff et al., HERMES Collaboration, Phys. Rev. Lett. **81**, 5519 (1998).
2. A. W. Thomas, Phys. Lett. B **126**, 97 (1983).
3. F.-G. Cao and A. I. Signal, Eur. Phys. J. C **21**, 105 (2001); Phys. Rev. D **68**, 074002 (2003).
4. F.-G. Cao and A. I. Signal, Phys. Rev. C **62**, 015203 (2000).
5. F.-G. Cao and A. I. Signal, Phys. Rev. D **60**, 074021 (1999); Phys. Lett. B **559**, 229 (2003).
6. H. Chen, F.-G. Cao and A. I. Signal, J. Phys. G **37**, 105006 (2010); G.-Q. Feng, F.-G. Cao, X.-H. Guo and A. I. Signal, Eur. Phys. J. C **72**, 2250 (2012).
7. A. Airapetian et al., HERMES Collaboration, Phys. Rev. Lett. **92**, 012005 (2004).

The predictions of $\psi(nS)$ and $\Upsilon(nS)$ decays based on the molecular structure

L. R. Dai[1,2,*] and E. Oset[2]

[1] *Department of Physics, Liaoning Normal University, Dalian, 116029, China*
[2] *Departamento de Física Teórica and IFIC, Centro Mixto Universidad de Valencia-CSIC, Institutos de Investigación de Paterna, Aptdo. 22085, 46071, Valencia, spain*
* E-mail: dailr@lnu.edu.cn

Firstly the previous results of J/ψ hadronic and radiative decays are reported, based on the vector-vector molecular structure of some tensor resonance states $f_2(1270)$, $f_2'(1525)$, $\bar{K}_2^{*0}(1430)$, $f_0(1370)$ and $f_0(1710)$ which are generated dynamically from the interaction of pairs of more elementary hadrons, finding the results reasonably consistent with the experimental data available. Then some new results are further reported for $\psi(nS)$ and $\Upsilon(nS)$ decays, including the $\psi(2S)$ decay into $\omega(\phi)f_2(1270)$, $\omega(\phi)f_2'(1525)$, $K^{*0}(892)\bar{K}_2^{*0}(1430)$ and radiative decay of $\Upsilon(1S)$, $\Upsilon(2S)$, $\psi(2S)$ into $\gamma f_2(1270)$, $\gamma f_2'(1525)$, $\gamma f_0(1370)$, $\gamma f_0(1710)$. Agreement with experimental data is found, and some predictions are done for future experiments, hoping to set further tests on the molecular nature of these resonances.

Keywords: Hadron interaction; dynamically generated resonances.

1. Motivation

The study of J/ψ decays has turned out a good tool to study the nature of some resonances, in particular those generated dynamically from the interaction of pairs of more elementary hadrons. Pioneering work along this line was done in the study of J/ψ decay into ϕ or ω and a scalar meson $\sigma(600)$ and $f_0(980)$.[1–4] Now we present some recent results of J/ψ hadronic decays[5,6] and radiative decays,[7] and also some new results of $\psi(nS)$ and $\Upsilon(nS)$ hadronic and radiative decays[8] to further test the molecular nature of $f_2(1270)$, $f_2'(1525)$, $\bar{K}_2^{*0}(1430)$, $f_0(1370)$, $f_0(1710)$ as dynamically generated states of vector meson-vector meson interaction within chiral unitary approach.[9]

2. The J/ψ Hadronic and Radiative Decays

In Ref. 5 we extended the idea[1-4] to study the J/ψ decay into ϕ, ω or $K^*(892)$ and some tensor resonances like the $f_2(1270)$, $f_2'(1525)$ and $K_2^*(1430)$, finding that the results are in good agreement with experiment. The idea is that the J/ψ decay proceeds via decay into a ϕ, ω or $K^*(892)$ that acts as spectator in the reaction and another pair of vector mesons in a certain invariant mass region that interact to create the $f_2(1270)$, $f_2'(1525)$ and $K_2^*(1430)$ resonances which are dynamically generated by the interaction of pairs of vectors. These pair of vector mesons rescatter after the primary production, producing the resonances that can be observed in the invariant mass distributions.[9] And another source of support for this idea stems from the radiative J/ψ decay into $f_2(1270)$, $f_2'(1525)$, $f_0(1370)$ and $f_0(1710)$.[7] In this case the photon is radiated from the charmed quarks and a pair of vectors is primary formed, leading through rescattering to the formation of these resonances. We define these ratios for J/ψ decays:

$$R_1 = \frac{\Gamma_{J/\psi \to \phi f_2(1270)}}{\Gamma_{J/\psi \to \phi f_2'(1525)}}; \qquad R_2 = \frac{\Gamma_{J/\psi \to \omega f_2(1270)}}{\Gamma_{J/\psi \to \omega f_2'(1525)}}; \ R_3 = \frac{\Gamma_{J/\psi \to \omega f_2(1270)}}{\Gamma_{J/\psi \to \phi f_2(1270)}},$$
$$R_4 = \frac{\Gamma_{J/\psi \to K^{*0} \bar{K}_2^{*0}(1430)}}{\Gamma_{J/\psi \to \omega f_2(1270)}}; \ R_T = \frac{\Gamma_{J/\psi \to \gamma f_2(1270)}}{\Gamma_{J/\psi \to \gamma f_2'(1525)}}; \ R_S = \frac{\Gamma_{J/\psi \to \gamma f_0(1370)}}{\Gamma_{J/\psi \to \gamma f_0(1710)}}. \tag{1}$$

The theoretical values of these ratios in Eq.(1) with their uncertainties are given in Table 1, which are taken from Ref. 5 for hadronic decays, and from Ref. 7 for radiative decays, respectively. We also list the experimental data with their uncertainties. It is seen that the theoretical results are in good agreement with experiment data available.

Table 1. Comparison between the experimental and the theoretical results of J/ψ hadronic decays and radiative decay.

	Theory	Experiment
	Hadronic decays[5]	
R_1	0.13 - 0.61 $(0.28^{+0.33}_{-0.15})$	0.22 - 0.47 $(0.33^{+0.14}_{-0.11})$
R_2	2.92 - 13.58 $(5.88^{+7.70}_{-2.96})$	12.33 - 49.00 $(21.50^{+27.50}_{-9.17})$
R_3	6.18 - 19.15 $(10.63^{+8.52}_{-4.45})$	11.21 - 23.08 $(15.85^{+7.23}_{-4.65})$
R_4	0.83 - 2.10 $(1.33^{+0.77}_{-0.50})$	0.55 - 0.89 $(0.70^{+0.19}_{-0.15})$
	Radiative decays[7]	
R_T	2 ± 1	$3.18^{+0.58}_{-0.64}$
R_S	1.2 ± 0.3	

3. Extension to the $\psi(nS)$ and $\Upsilon(nS)$ Hadronic and Radiative Decays

Since the arguments used in the J/ψ decays are very simple and general and can be easily extended to decays of $\psi(nS)$ and $\Upsilon(nS)$. Taking advantage of the fact that some of these decays have become recently available, we hope to extend the J/ψ decays to $\psi(nS)$ and $\Upsilon(nS)$ hadronic and radiative decays to further test on the molecular nature. Now the ratios are defined:

$$
\begin{aligned}
&\widetilde{R_1} = \frac{\Gamma_{\psi(2S)\to\phi f_2(1270)}}{\Gamma_{\psi(2S)\to\phi f_2'(1525)}}; \quad
\widetilde{R_2} = \frac{\Gamma_{\psi(2S)\to\omega f_2(1270)}}{\Gamma_{\psi(2S)\to\phi f_2(1270)}}; \quad
\widetilde{R_3} = \frac{\Gamma_{\psi(2S)\to K^{*0}K_2^{*0}(1430)}}{\Gamma_{\psi(2S)\to\omega f_2(1270)}}, \\
&\widetilde{R_T} = \frac{\Gamma_{\Upsilon(1S)\to\gamma f_2(1270)}}{\Gamma_{\Upsilon(1S)\to\gamma f_2'(1525)}}; \quad
\widetilde{R_S} = \frac{\Gamma_{\Upsilon(1S)\to\gamma f_0(1370)}}{\Gamma_{\Upsilon(1S)\to\gamma f_0(1710)}}; \quad
\widehat{R_T} = \frac{\Gamma_{\psi(2S)\to\gamma f_2(1270)}}{\Gamma_{\psi(2S)\to\gamma f_2'(1525)}}, \quad (2) \\
&\widehat{R_S} = \frac{\Gamma_{\psi(2S)\to\gamma f_0(1370)}}{\Gamma_{\psi(2S)\to\gamma f_0(1710)}}; \quad
\overline{R_T} = \frac{\Gamma_{\Upsilon(2S)\to\gamma f_2(1270)}}{\Gamma_{\Upsilon(2S)\to\gamma f_2'(1525)}}; \quad
\overline{R_S} = \frac{\Gamma_{\Upsilon(2S)\to\gamma f_0(1370)}}{\Gamma_{\Upsilon(2S)\to\gamma f_0(1710)}}.
\end{aligned}
$$

The theoretical values of these ratios[8] in Eq.(2) with their uncertainties for the $\psi(nS)$ and $\Upsilon(nS)$ hadronic and radiative decays are given in Table 2, in which the experimental data with their uncertainties are also listed. It is seen that the agreement of the results with the data available on $\Upsilon(1S)$ to $\gamma\, f_2(1270)$ and $\gamma\, f_2'(1525)$ is good and the same occurs with the decay of $\psi(2S)$ into $\omega\, f_2(1270)$, $\phi\, f_2'(1525)$ and $K^*(892)\, K_2^*(1430)$. We also make predictions for other related decays not yet available.

Table 2. Comparison between the experimental and the theoretical results of $\psi(2S)$ decays. $\widetilde{R_1} \cdot \widetilde{R_2}$ provides the ratio $\Gamma_{\psi(2S)\to\omega f_2(1270)}/\Gamma_{\psi(2S)\to\phi f_2'(1525)}$. Experimental results for $\widetilde{R_3}$ and $\widetilde{R_1}\cdot\widetilde{R_2}$ from Refs. 10–12, and radiative data from Refs. 13–15.

	Theory	Experiment
	Hadronic decays	
$\widetilde{R_1}$	0.12-0.56 $(0.26^{+0.30}_{-0.14})$	
$\widetilde{R_2}$	5.91-18.30 $(10.16^{+8.14}_{-4.25})$	
$\widetilde{R_3}$	0.88-2.22 $(1.41^{+0.82}_{-0.53})$	0.54-1.33 $(0.86^{+0.47}_{-0.32})$
$\widetilde{R_1} \cdot \widetilde{R_2}$	1.24-5.74 $(2.64^{+3.1}_{-1.4})$	3.0-9.3 $(5.0^{+4.3}_{-2.0})$
	Radiative decays	
$\widetilde{R_T}$ $(\Upsilon(1S))$	1.84 ± 0.92	$2.66^{+1.13}_{-0.70}$
$\widetilde{R_S}$ $(\Upsilon(1S))$	1.05 ± 0.26	
$\widehat{R_T}$ $(\psi(2S))$	1.94 ± 0.97	
$\widehat{R_S}$ $(\psi(2S))$	1.14 ± 0.28	
$\overline{R_T}$ $(\Upsilon(2S))$	1.83 ± 0.92	
$\overline{R_S}$ $(\Upsilon(2S))$	1.05 ± 0.26	

4. Conclusions

In the present work we show the results on the test of the molecular nature of the $f_2(1270)$, $f'_2(1525)$, $K^*_2(1430)$, $f_0(1370)$ and $f_0(1710)$ resonances by investigating the hadronic decays and radiative decays of $\psi(nS)$ and $\Upsilon(nS)$ states. The results are in agreement with experimental data available. Predictions are also done for ratios not yet measured, hoping future experiments to set further tests on the molecular nature of these resonances.

Acknowledgments

Supported by Natural Science Foundation of the Liaoning Scientific Committee (2013020091) and National Natural Science Foundation of China (11375080,10975068), and supported by the Spanish Ministerio de Economiay Competitividad and European FEDER funds under the contract number FIS2011-28853-C02-01, and the Generalitat Valenciana in the program Prometeo, 2009/090. We acknowledge the support of the European Community-Research Infrastructure Integrating Activity Study of Strongly Interacting Matter (acronym HadronPhysics3, Grant Agreement no. 283286) under the Seventh Framework Programme of EU.

References

1. U. -G. Meissner and J. A. Oller, Nucl. Phys. A **679**, 671 (2001).
2. L. Roca, J. E. Palomar, E. Oset and H. C. Chiang, Nucl. Phys. A **744**, 127 (2004).
3. T. A. Lahde and U. -G. Meissner, Phys. Rev. D **74**, 034021 (2006).
4. B. Liu, M. Buescher, F. -K. Guo, C. Hanhart and U. -G. Meissner, Eur. Phys. J. C **63**, 93 (2009).
5. A. Martinez Torres, L. S. Geng, L. R. Dai, B. X. Sun, E. Oset and B. S. Zou, *Phys. Lett. B* **680**, 310 (2009).
6. L. R. Dai, *AIP Conf. Proc.***1322**, 405 (2010).
7. L. S. Geng, F. K. Guo, C. Hanhart, R. Molina, E. Oset and B. S. Zou, *Eur. Phys. J. A* **44**, 305 (2010).
8. Lianrong Dai and Eulogio Oset, *Eur. Phys .J. A* **49**, 130 (2013).
9. L. S. Geng and E. Oset, *Phys. Rev. D* **79**, 074009 (2009).
10. J. Beringer *et al.* [PDG], *Phys. Rev. D* **86**, 010001 (2012).
11. J. Z. Bai *et al.* (BES Collaboration), *Phys. Rev. D* **69**, 072001 (2004).
12. J. Z. Bai et al. (BES Collaboration), *Phys. Rev. Lett.* **81**, 5080 (1998).
13. D. Besson et al. CLEO Collaboration, *Phys. Rev. D* **75**, 072001 (2007).
14. D. Besson et al. (CLEO Collaboration), *Phys. Rev. D* **83**, 037101(2011).
15. S. B. Athar et al. CLEO Collaboration, *Phys. Rev. D* **73**, 032001(2006).

How to distinguish a molecule from an 'elementary' particle?

L. Y. Dai[1]*, M. Shi[1], G. Y. Tang[1], H. Q. Zheng[1,2]

1) Department of Physics and State Key Laboratory of Nuclear Physics and Technology, Peking University, Beijing 100871, P. R. China
2) Collaborative Innovation Center of Quantum Matter, Beijing, P. R. China

We discuss how to use Morgan's pole counting rule to distinguish a molecular state from an 'elementary' particle. As two examples we focus on $X(3872)$ and $f_0(980)$ particles. A molecule may be generated from a meson loop bubble chain, and an 'elementary' particle is related to an explicit interaction field in the effective lagrangian and propagates with a Breit–Wigner propagator. For $X(3872)$ it is found that the data favor the 'elementary' particle explanation. For $f_0(980)$ the study becomes much more difficult, since highly non-perturbative dynamics is involved. A unitarization model analysis suggests that $f_0(980)$'s property is quite exotic. Unlike other light scalars, it does not behave like a $\bar{q}q$ state, and could be interpreted as a molecule.

Keywords: Hadronic molecule; chiral symmetry; X(3872); $f_0(980)$.

1. A Brief Introduction to the Pole Counting Rule

It has been a longstanding and challenging problem for particle physicists to judge wether a particle is elementary or composite, from experimentally known cross-section or phase shift data. Unfortunately, the exact definition of a 'molecule' and an 'elementary' particle in hadron physics is not totally clear in the context of quantum field theory. Hence the topic of this talk can only be clear in quantum mechanics or quantum scattering theory.

In the context of quantum scattering theory, it is pointed out long time ago that,[1] in some special cases, it is possible to solve the problem model independently, by counting the number of poles near a threshold, in an s-wave amplitude. To see this, notice that any scattering amplitude can be

---

*Present address: Theory Center, Thomas Jefferson National Accelerator Facility, Newport News, VA 23606, USA.

written as,

$$T = \frac{1}{M - ik} \tag{1}$$

and (resonance) poles correspond to the zeros of $M - ik$, and M can be expanded in powers of k^2,

$$M(k^2) = -\frac{1}{a} + \frac{1}{2} r_{\text{eff}} k^2 + O(k^4) , \tag{2}$$

where a is the scattering length, r_{eff} the effective range parameter. Typically r_{eff} is of the order of strong interaction range and be small, hence $M - ik$ only contains one zero near threshold when the scattering length a is large.

On the other hand when there is a CDD pole weakly couples to this channel, there will be two poles near the threshold, $M(k^2)$ can be expressed as,

$$M(k^2) = \frac{k^2 - k_0^2}{g^2} + \text{rescattering corrections} \tag{3}$$

where g is small. In this situation, there appears a pair of poles on the k plane, which implies the occurrence of an elementary particle.[1]

The above explanation is confined to the single channel situation but the pole counting method also works in the couple channel situation.[1] Applying the pole counting method, it is found that $X(3872)$ contains two poles near threshold.[2] For this reason it is argued that $X(3872)$ contains large $c\bar{c}$ component. In Ref. 2 it is assumed that $X(3872)$ propagates through a Breit–Wigner propagator,

$$D(E) = E - E_f + \frac{i}{2}(g_1 k_1 + g_2 k_2 + \Gamma(E) + \Gamma_c) , \tag{4}$$

where subscripts 1, 2 imply $D^0 \bar{D}^{0*}$ channel and $D^+ D^{*-}$ channel respectively; $\Gamma(E)$ denotes the partial decay widths into $J/\Psi\rho$ and $J/\Psi\omega$; and Γ_c denotes decay width into all other channels except the two mentioned above.

A Breit–Wigner propagator has two built-in poles. However, if the data prefers a one-pole description, the parameters chosen by the fit will naturally push one pole away from the threshold. Through the fit it is found that two nearby poles are needed in order to describe the data, if the branching ratio $\text{Br}(B \to KX)$ be reasonably small. Hence, according to the pole counting rule, the $X(3872)$ particle is of $\bar{c}c$ origin, though it ought to be heavily renormalized by the $\bar{D}D^{*0}$ cloud since it is very close to the threshold. Similar conclusion is also obtained in Ref. 3 in a different approach.

It is worthwhile mentioning that the above approach may still be somewhat model dependent. For example one may question whether the conclusion may depend on the assumption that a Breit–Wigner type propagator is used to describe $X(3872)$. Could it happen that a bubble chain description fits the data equally well? To answer this question we extend the previous study to a more general case by considering the two mechanism simultaneously in the next section.[4]

2. Reconsideration of $X(3872)$

Here we extend the discussion made in Ref. 2 by including both the $X(3872)$ propagator and the DD^* bubble chain. The appropriate effective lagrangian may be written as the following:

$$
\begin{aligned}
\mathcal{L}_{D\bar{D}^*} &= \lambda_1(\bar{D}^{*\mu}DD^{*\mu}D + \bar{D}D^{*\mu}\bar{D}D^{*\mu}) - 2\lambda_1(\bar{D}^{*\mu}D\bar{D}D^{*\mu}), \\
\mathcal{L}_{XD\bar{D}^*} &= g_1 X^{\mu}(\bar{D}D_{\mu}^* - \bar{D}_{\mu}^*D), \\
\mathcal{L}_{BXK} &= ig_2 X^{\mu}(\bar{B}\partial_{\mu}K + \text{h.c.}), \\
\mathcal{L}_{BKD\bar{D}^*} &= ig_3(\bar{D}D_{\mu}^* - \bar{D}_{\mu}^*D)(\bar{B}\partial^{\mu}K + \text{h.c.}),
\end{aligned} \tag{5}
$$

and the Feynman diagrams are depicted in Figs. 1 and 2. Here the X particle resembles a $\bar{c}c$ state and the DD^* bubble chain will be expected to generate a $\bar{D}D^*$ molecule and we let data determine which mechanism is preferred. The decay diagrams in Fig. 2 can be summed up[5] and the amplitude is used in the data fitting. Unlike Ref. 2, here the situation becomes more complicated since there exists a competition between the Breit–Wigner propagator and the molecular bubble chain. Though the result relies on inaccurate data, one can still conclude, in any case, that $\bar{c}c$ component plays a important and non-negligible role, in agreement with Ref. 2. For more details we refer to Ref. 4.

3. On the Property of $f_0(980)$

The above application for the pole counting rule is a successful one. Here we turn to discuss the property of $f_0(980)$, which is much more difficult and confused than the previous example. Because strong non-perturbative dynamics is involved and the data on phase shifts and inelasticity around $\bar{K}K$ threshold are inaccurate.[6] The original pole counting method[1] only works for weak interaction case, i.e., it only involves weakly bounded state or weakly coupled CDD pole. In such a situation, it is impossible to judge whether a distant pole, if exist, physically relates to certain threshold. This

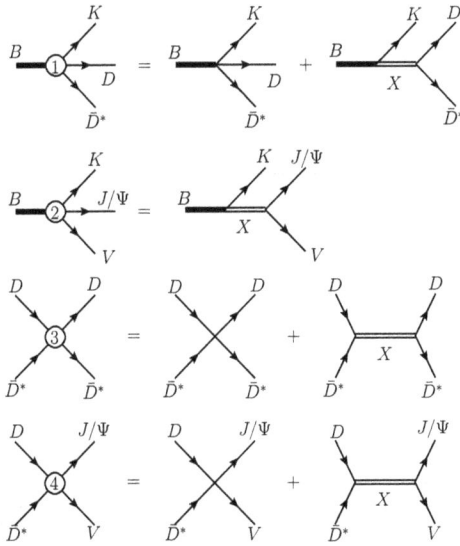

Fig. 1.   The effective vertices.

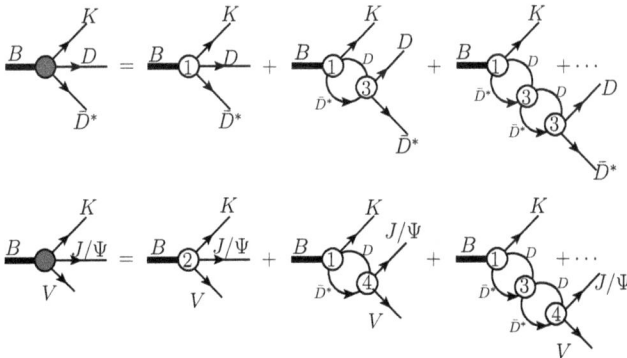

Fig. 2.   The complete B decay chains. Notice inside which both the bubble chain and the (dressed) $X(3872)$ are involved.

situation may happen when the interaction becomes strong. An example is the $\pi\pi$, $\bar{K}K$ couple channel system, there one may raise questions such as whether the broad $f_0(600)$ be a $\pi\pi$ molecule? or whether the $f_0(980)$ pole is a $\bar{K}K$ molecule?

In Ref. 7 the number of poles around $\bar{K}K$ threshold is studied and indeed two nearby poles are found. Nevertheless it is noticed that the third sheet pole position (see Table 1) is very unstable and can be easily tuned away without sacrificing the quality of the fit.

Table 1. $f_0(980)$ pole position and residue in different channels.

Pole position	$g_{\pi\pi}^2$	$g_{\bar{K}K}^2$
$\sqrt{s_{II}} = 0.999 - 0.021i$	$-0.07 - 0.01i$	$-0.10 + 0.09i$
$\sqrt{s_{III}} = 0.977 - 0.060i$	$-0.10 + 0.02i$	$-0.02 - 0.09i$

Beside the instability of the (third sheet) pole position, the $f_0(980)$ has a tiny di-photon width (hence not a $\bar{q}q$ state), see Table 2. Meanwhile,

Table 2.   Di-photon width of the $f_0(980)$ pole.

	Pole positions(GeV)	$\Gamma(f_J \to \gamma\gamma)$(keV)
$f_0^{II}(980)$	$0.999 - 0.021i$	0.12
$f_0^{III}(980)$	$0.977 - 0.060i$	0.35

the $f_0(980)$ particle also maintains some odd properties which is hard to understand: it has a negative (the real part) residue when couples to $\pi\pi$ and $\bar{K}K$ (see Table 1).

On the contrary, the broad $f_0(600)$ also has a negative real part of the residue, but that seems to be well understood,[8] because $f_0(600)$ is broad and it dominates the low energy phase shift, its residue has to be negative. This observation is obtained by making use of the production representation for partial wave amplitudes.[9]

Alternative approach is also explored in the literature. In Ref. 10 the $f_0(980)$ particle is studied in the extended Nambu–Jona-Lasinio model. Using heat kernel expansion technique, one obtains an effective chiral lagrangian with a scalar nonet (i.e., in a linear realization of chiral symmetry). Here scalars are chiral partners of Nambu-Goldstone bosons. Scattering amplitudes obtained by a K-matrix unitarization are analyzed. It is found that with an unnaturally small cutoff parameter Λ, and a bare $M_\sigma = 1$ GeV, one gets the mass and width of σ ($f_0(600)$), κ ($K^*(700)$) and $a_0(980)$ simultaneously, except the $f_0(980)$. This observation strongly suggests the peculiarity of $f_0(980)$.

The situation as discussed above forces one to go beyond the original pole counting rule of Ref. 1 for the purpose of investigating $f_0(980)$. In the following we propose the study through using the unitarized χPT amplitudes and the large N_c technique.[11] In Ref. 12 a careful study is made on the pole structure of $\pi\pi - K\bar{K}$ and $\pi\eta - K\bar{K}$ couple channel [1,1] matrix Padé amplitudes of $SU(3) \times SU(3)$ chiral perturbation theory. By fitting phase shift and inelasticity data, pole positions in different channels ($f_0(980)$, $a_0(980)$, $f_0(600)$, $K_0^*(800)$, $K^*(892)$, $\rho(770)$) are determined and their N_c trajectories are traced. We stress that a couple channel Breit–Wigner resonance should exhibit two poles on different Riemann sheets that reach the same position on the real axis when $N_c = \infty$. Poles are hence classified using this criteria and we conclude that $K^*(892)$ and $\rho(770)$ are unambiguous Breit–Wigner resonances. For scalars the situation is much less clear. We find that there is no shadow pole accompanied by $f_0(980)$, while $f_0(600)$ and $a_0(980)$, though behave oddly when varying N_c, do maintain a twin pole structure. Hence through this analysis we conclude that the $f_0(980)$ is not of a Breit–Wigner type and is therefore may be a molecular state.

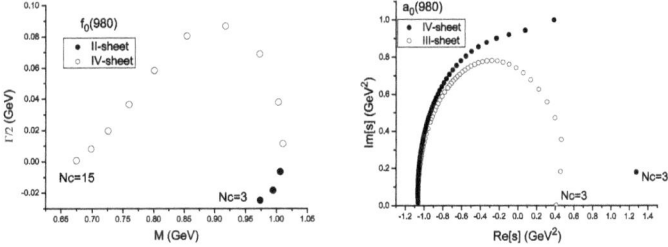

Fig. 3. N_c trajectories for $f_0(980)$ and $a_0(980)$.

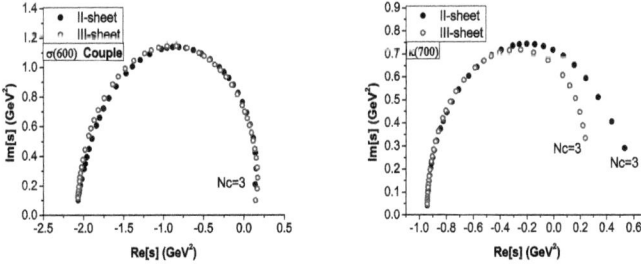

Fig. 4. N_c trajectories for $f_0(600)$ and $K(700)$, in the matrix Padé amplitude.

To summarize the discussion on $f_0(980)$: It does not behave like a conventional $q\bar{q}$, through its di-photon decay width analysis. It should neither be a tetra quark. Since in the present scheme, we believe that a 'tetra-quark' is also elementary rather than molecular, hence it's of Breit–Wigner type and should also maintain a twin pole structure. Furthermore, it seems not to be in a chiral multiplet.[10] Hence the $f_0(980)$ prefers a molecule interpretation. At last, it is worth emphasizing that the properties of $f_0(980)$ as discussed above is in nice agreement, at qualitative level, with the most recent and careful analysis of Dai and Pennington,[13] which for the first time takes the $\bar{K}K$ data into account when determining the the the di-photon width.

Acknowledgments

I would like to thank the organizers of *Chiral* 2013, especially Li-sheng Geng for kind hospitality and providing the nice atmosphere during the conference. This work is supported in part by National Nature Science Foundations of China under Contract Nos. 10925522 and 11021092.

References

1. D. Morgan, Nucl. Phys. A543(1992)632.
2. O. Zhang, C. Meng, H. Q. Zheng, Phys. Lett. B680 (2009)453.
3. Yu. Kalashnikova, A. V. Nefediev, Phys. Rev. D80(2009)074004.
4. C. Meng et al., to appear.
5. Similar techniques were used in, e.g., N. N. Achasov and G. N. Shestakov, Phys. Rev. D86(2012)114013; G. Y. Chen, Q. Zhao, Phys. Lett. B718(2013)1369.
6. For an old review of literature on $f_0(980)$, see, M. P. Locher, V. E. Markushin, H. Q. Zheng, Eur. Phys. J. C4 (1998) 317.
7. Y. Mao et al., Phys. Rev. D79 (2009) 116008.
8. X. G. Wang et al., PoS CD09 (2009) 084; talk given by H.Q. Zheng at *Chiral Dynamics* 09, 6-10 Jul 2009, Bern, Switzerland.
9. H. Q. Zheng et al., Nucl.Phys. A733 (2004) 235; Z. Y. Zhou et al., JHEP 0502 (2005) 043; Z. Y. Zhou, H. Q. Zheng, Nucl.Phys. A775 (2006) 212. For a review, see, H. Q. Zheng, Front. Phys. China. 8 (2013) 540.
10. M. X. Su et al., Nucl. Phys. A792 (2007) 288.
11. This method has been studied in the literature, for example, J. R. Pelaez, G. Rios, Phys. Rev. Lett. 97 (2006) 242002; Z. X. Sun et al., Mod. Phys. Lett. A22(2007) 711; Z. H. Guo, J. A. Oller, Phys.Rev. D84 (2011) 034005.
12. L. Y. Dai et al., Commun. Theor. Phys. 58 (2012) 410; Commun. Theor. Phys. 57 (2012) 841.
13. L. Y. Dai, M. R. Pennington, arXiv:1403.7514 [hep-ph]

A chiral effective approach to parity-odd proton-proton scattering

J. de Vries

*Institute for Advanced Simulation, Institut für Kernphysik, and
Jülich Center for Hadron Physics, Forschungszentrum Jülich,
D-52425 Jülich, Germany
E-mail: j.de.vries@fz-juelich.de*

These proceedings summarize recent work on parity violation in proton-proton scattering in chiral effective field theory.

Keywords: Parity violation; chiral effective field theory.

1. Parity Violation in Hadronic Systems

The discovery in 1957 of parity violation (PV) in weak interactions came as a surprise, but turned out to be one of the key ingredients in unraveling the structure of the Standard Model of particle physics. It is now understood that at the fundamental level, PV arises from the exchange of the heavy weak gauge bosons which, at low energies, gives rise to PV four-fermion interactions. Exchange of the charged weak bosons induces the beta-decay of the muon, neutron, and certain atomic nuclei. Since no electromagnetic or strong counterpart of such a process exists, the charged weak current is relatively easily experimentally identified. This is in stark contrast with the effects of neutral weak gauge boson exchange in hadronic systems. In this case, the effects of the PV four-quark operators are swamped by the much larger strong and electromagnetic interactions.

Despite this difficulty several experiments have successfully observed PV due to the neutral weak interaction in hadronic systems (see Ref. 1 for a review). In particular, there exist several data points for the PV longitudinal analyzing power (LAP) in proton-proton (pp) scattering. Enhancement factors due to nuclear structure have led to experiments on more complicated systems such as ^{19}F and ^{133}Cs, but such systems suffer from larger theoretical uncertainties.

On the theoretical side the challenge is to describe the experimental patterns in terms of the PV four-quark operators. Because of confinement, the latter do not appear directly, but, instead, induce effective PV interactions among hadrons. The most important ones being the interactions among the nucleon and the pion, although heavier hadrons can play an important role as well. Historically, the PV nucleon-nucleon potential has been described in terms of one-meson exchanges involving such PV nucleon-meson interactions. This potential is often called the DDH-potential after the authors of Ref. 2.

More recent work has approached the problem using effective field theory (EFT) techniques. At very low energies all hadrons, apart from the external nucleons, can be integrated out and PV can be described in terms of five nucleon-nucleon contact interactions (one for each possible S- to P-wave transition). It can been shown that at sufficiently low energies, the DDH-potential and the so-called pionless EFT are identical modulo higher-order effects due to higher partial-wave transitions.[1,3]

Some of the PV observables involve significant momentum scales (in the form of relatively large external energies or binding energies), which makes it necessary to include pions explicitly. This has lead to the derivation of the 'pionfull' effective PV potential. At leading order (LO) this potential consists of a one-pion-exchange (OPE) term involving the weak pion-nucleon coupling constant h_π.[4] At next-to-leading order (NLO) the same five contact interactions appear as in the pionless EFT, but, additionally, two-pion-exchange (TPE) diagrams start contributing.[5]

The PV pionfull EFT has the major advantage that it can be naturally combined with its very successful P-even counterpart. The P-conserving nucleon-nucleon potential has been derived up to next-to-next-to-next-to-leading order (N³LO) and describes the nucleon-nucleon scattering data-base with a reduced χ^2 close to unity.[6] In very recent work, we have combined the P-even and P-odd potential in a consistent way and used the resulting potential to investigate PV in pp scattering.[7]

2. Ingredients and Formalism

The LO PV potential arises from an OPE diagram and is proportional to the isospin structure $(\vec{\tau}_1 \times \vec{\tau}_2)^3$, where $\vec{\tau}_i$ denotes the isospin of the scattered nucleons. It is not hard to see that this structure vanishes between states of total equal isospin and does not contribute to pp scattering. This property has lead most earlier analyses of the pp LAP to not take the weak pion-nucleon coupling h_π into account.[8] However, in the modern EFT picture, h_π contributes at higher orders.

At NLO the potential consists of TPE diagrams and five nucleon-nucleon contact interactions. In case of pp scattering, the potential can be compactly written as

$$V_{\rm PV} = \left[i(\vec{\tau}_1 + \vec{\tau}_2)^3 (\vec{\sigma}_1 \times \vec{\sigma}_2) \cdot \vec{q}\right] \left[h_\pi f(\vec{q}) + \frac{C}{2F_\pi \Lambda_\chi^2}\right], \quad (1)$$

in terms of the exchanged momentum $\vec{q} = \vec{p} - \vec{p}'$ between nucleon 1 and 2 and their spin $\vec{\sigma}_i$. At this order, h_π does contribute and is associated with the complicated TPE function $f(\vec{q})$ (for the explicit expression, see Ref. 7), while C is the low-energy constant associated with the nucleon-nucleon contact interactions (the factor $1/(2F_\pi \Lambda_\chi^2)$ with $F_\pi = 92.4$ MeV and $\Lambda_\chi = 1$ GeV is inserted to make C dimensionless).

We combine the above PV potential with the N^{3}LO chiral potential[6] to obtain the total potential V. The Lippmann-Schwinger (LS) equation is solved to obtain the T-matrix in the presence of PV. The integral over intermediates states in the LS equation is divergent and we regularize the total potential by a cut-off function. We vary the cut-off between 450 and 600 MeV to estimate the theoretical uncertainty of our results. Finally, with the solution of the T-matrix, it is straightforward to obtain the scattering amplitude and the scattering observables.

3. Results

The observable we are interested in is the LAP in pp scattering which vanishes in the limit of P conservation. It is defined as the difference in cross section between an unpolarized target and a beam of positive and negative helicity, normalized to the sum of these cross sections. The LAP has been measured for several beam energies and results for reported for 13.6, 45, and 221 MeV (lab energy).[9]

We fit the coupling constants h_π and C appearing in Eq. (1) to the three data points. At the level of a total $\chi^2 = 2.71$ we obtain the following allowed ranges for the coupling constants

$$h_\pi = (1.1 \pm 2) \cdot 10^{-6}, \quad (2)$$
$$C = (-6.5 \pm 8) \cdot 10^{-6}, \quad (3)$$

and the couplings are heavily correlated (see Ref. 7 for more details). The large uncertainty is dominated by experimental errors and the lack of data points. The cut-off dependence of the fits play a much smaller role. Additional experiments are needed to reduce the uncertainties in the fits.

It is interesting to compare our allowed range for the pion-nucleon coupling h_π with theoretical estimates and experimental bounds. The authors of Ref. 2 obtained the range $0 \leq h_\pi \leq 1.2 \cdot 10^{-6}$ with a 'best' value of $h_\pi = 4.6 \cdot 10^{-7}$ well within the range of our fit. In Ref. 10 a smaller estimate $h_\pi \simeq 1 \cdot 10^{-7}$ was obtained. This smaller estimate is in agreement with a direct lattice calculation[11] of $h_\pi = (1.1 \pm 0.5\,(\text{stat}) \pm 0.5\,(\text{sys})) \cdot 10^{-7}$ and the experimental limit obtained from ^{19}F γ-ray emission $h_\pi < 1.3 \cdot 10^{-7}$,[12] but the latter bound might have uncontrolled theoretical uncertainties. Regardless, our fit results in Eq. (2) are consistent with the smaller estimates and the experimental bound, but the uncertainties are currently too large to say much more.

The analysis of the LAP in pp scattering is only the first step in analyzing hadronic parity violation in our EFT framework. At the moment, work is in process to derive the parity-odd potential up to higher orders and to study different processes such as $np \to d\vec{\gamma}$. Furthermore, we wish to extend our EFT calculations to systems with more than two nucleons.

Acknowledgments

I would like to thank Ulf-G. Meißner, E. Epelbaum, and N. Kaiser for the enjoyable collaboration on the work presented here. I am grateful to the organizers of the conference for the invitation and the wonderful time in Beijing. This work is supported in part by DFG and NSFC (CRC 110).

References

1. W. C. Haxton and B. R. Holstein, Prog. Part. Nucl. Phys. **71** (2013) 185.
2. B. Desplanques, J. F. Donoghue and B. R. Holstein, Annals Phys. **124** (1980) 449.
3. M. R. Schindler and R. P. Springer, Prog. Part. Nucl. Phys. **72** (2013) 1.
4. D. B. Kaplan and M. J. Savage, Nucl. Phys. A **556** (1993) 653.
5. S.-L. Zhu et al., Nucl. Phys. A **748** (2005) 435.
6. E. Epelbaum, W. Glöckle and U.-G. Meißner, Nucl. Phys. A **747** (2005) 362.
7. J. de Vries, U.-G. Meißner, E. Epelbaum and N. Kaiser, Eur. Phys. J. A **49** (2013) 149.
8. J. Carlson, R. Schiavilla, V. R. Brown and B. F. Gibson, Phys. Rev. C **65** (2002) 035502.
9. P. D. Eversheim et al., Phys. Lett. B **256** (1991) 11; S. Kistryn et al., Phys. Rev. Lett. **58** (1987) 1616; A. R. Berdoz et al. [TRIUMF E497 Collaboration], Phys. Rev. Lett. **87** (2001) 272301.
10. U.-G. Meißner and H. Weigel, Phys. Lett. B **447** (1999) 1.
11. J. Wasem, Phys. Rev. C **85** (2012) 022501.
12. S. A. Page et al., Phys. Rev. C **35** (1987) 1119.

$\bar{K}N$ reactions and baryon resonances with strangeness -1

Zhi-Hui Guo

Department of Physics, Hebei Normal University, 050024 Shijiazhuang, P. R. China
E-mail: zhguo@mail.hebtu.edu.cn

J. A. Oller

Departamento de Física, Universidad de Murcia, E-30071 Murcia, Spain
E-mail: oller@um.es

We study the $\Lambda(1405)$, $\Lambda(1670)$ and the isovector baryon resonances of Σ-type in the pseudo-Goldstone octet and the lightest baryon octet scattering. S-wave scattering is considered by employing unitary chiral perturbation theory. Through fitting various experimental data, such as $\bar{K}N$ cross sections, $\pi\Sigma$ event distributions, atomic kaon hydrogen data, etc., we determine the free parameters, which are then used in the discussion of the baryon spectroscopy and the $\bar{K}N$ scattering lengths.

Keywords: $\Lambda(1405)$; $\bar{K}N$ scattering length; unitary chiral perturbation theory.

1. Introduction

Recent study of the $\Lambda(1405)$ resonance from the chiral effective theory reveals an interesting scenario in which this resonance corresponds to a two-pole structure in the complex energy plane.[1] This interesting picture has been pursued by many theoretical works.[2-5] It is worth pointing out that the two-pole structure is also experimentally verified by the CLAS Collaboration in the $\pi\Sigma$ photoproduction[6] and the $\Lambda(1405)$ line shape in electroproduction.[7]

In this work, we further investigate the $\Lambda(1405)$ properties by including the new results of the energy shift and width of the kaon hydrogen $1s$ state from the SIDDHARTA Collaboration,[8] which is measured with much smaller uncertainties than those from KEK[9] and DEAR[10] Collaborations. In addition, the $\Lambda(1670)$ and other isovector baryon resonances are studied in detail in the unitarized chiral perturbation theory.[11] We summarize the main findings of the latter reference in the following.

2. Theoretical Framework and Phenomenological Results

The key object in our study is two-body partial-wave amplitudes. To calculate it, we employ the algebraic solution of the N/D method

$$T(W) = [1 + N(W) \cdot g(W^2)]^{-1} \cdot N(W) , \qquad (1)$$

being W the center of mass (c.m.) energy. In this solution, the function $g(s)$, $s = W^2$, incorporates the right-hand or unitarity cuts from the s-channel two-particle intermediate states, while the $N(W)$ function only includes the cuts generated from the crossed channel dynamics. The function $g(W)$, as well as the perturbative set up to calculate $N(W)$ in Chiral Perturbation Theory, are given in detail in Refs. 1,2 and references therein. The ansatz for the N/D method in the previous equation can be easily extended to the coupled-channel situation and in that case $T(W), g(W), N(W)$ should be generalized to matrices.[1] We study the S-wave amplitudes and consider ten relevant channels: $\pi^0\Lambda$, $\pi^0\Sigma^0$, $\pi^-\Sigma^+$, $\pi^+\Sigma^-$, K^-p, $\bar{K}^0n$, $\eta\Lambda$, $\eta\Sigma^0$, $K^0\Xi^0$ and $K^+\Xi^-$.

Using the key object in Eq. (1), the cross section for the $(\phi B)_i \to (\phi B)_j$ scattering then takes the form

$$\sigma[(\phi B)_i \to (\phi B)_j] = \frac{1}{16\pi s} \frac{|\vec{p_j}|}{|\vec{p_i}|} |T_{(\phi B)_i \to (\phi B)_j}|^2 , \qquad (2)$$

with $\vec{p_i}$ and $\vec{p_j}$ the c.m. three momenta of the ingoing and outgoing baryons respectively. In the fit, we consider eight cross sections from various processes: $K^-p \to K^-p$, $K^-p \to \bar{K}^0n$, $K^-p \to \pi^+\Sigma^-$, $K^-p \to \pi^-\Sigma^+$, $K^-p \to \pi^0\Sigma^0$, $K^-p \to \pi^0\Lambda$, $K^-p \to \eta\Lambda$ and $K^-p \to \pi^0\pi^0\Sigma^0$. Two types of $\pi\Sigma$ event distributions are considered in the fit by introducing two more parameters r, r'.[11] One can express the energy shift and width of the kaon hydrogen state in terms of the $\bar{K}N$ scattering length, which can be found in detail in Ref. 12. In this way, we can include the new measurement from SIDDHARTA collaboration in our fit. The relation between the K^-p scattering length a_{K^-p} and the unitarized partial-wave amplitude in Eq. (1) takes the form

$$\mathcal{F}(\sqrt{s}) \equiv \frac{T_{K^-p \to K^-p}(\sqrt{s})}{8\pi\sqrt{s}} , \qquad a_{K^-p} = \mathcal{F}(M_{K^-} + m_p) . \qquad (3)$$

For other experimental inputs in the fit, we refer to Ref. 11 for details.

Two different strategies are used in the phenomenological fits. In Fit I, we do not discriminate the pseudo-Goldstone decay constants and take one common value for all the channels that is fitted to data. While in Fit II, we incorporate the physical values for the π, K, and η decay constants. In each case, we have also tried two definitions for χ^2, namely the standard one and another one in which the contribution of each observable to the previous χ^2 is divided by its number of experimental points.[11] Roughly speaking, we do not find significant variants in the reproduction of data when switching between Fit I and Fit II, which can both reasonably reproduce the experimental inputs, regardless of the definitions of the χ^2 employed.

The final results for the $\bar{K}N$ scattering lengths (in units of fm) after combining Fit I and Fit II are

$$a_{K^-p} = (-0.69 \pm 0.16) + i\,(0.94 \pm 0.11)\,,$$
$$a_{I=0} = (-1.74 \pm 0.34) + i\,(1.31 \pm 0.20)\,,$$
$$a_{I=1} = (0.36 \pm 0.12) + i\,(0.56 \pm 0.12)\,, \tag{4}$$

where the subscript I stands for the isospin and the results are in good agreement with recent determinations from Refs. 3–5.

For the spectroscopy of the baryon resonances, we have obtained their pole positions in the appropriate Riemann sheets both for isoscalar and isovector ones. We refer to Ref. 11 for the definitions of different Riemann sheets. The pole positions of the $\Lambda(1670)$ from Fit I and Fit II read

$$\sqrt{s} = (1674^{+3}_{-3} - i\,11^{+7}_{-3})\ \text{MeV}\,, \quad \text{(Fit I)}$$
$$\sqrt{s} = (1677^{+5}_{-3} - i\,11^{+5}_{-3})\ \text{MeV}\,, \quad \text{(Fit II)} \tag{5}$$

where only the resonance poles in the fifth Riemann sheet are shown. For the $\Lambda(1405)$, we confirm that it corresponds to a two-pole structure in the complex energy plane and their explicit values from both fits read

$$\sqrt{s_1} = (1436^{+14}_{-10} - i\,126^{+24}_{-28})\ \text{MeV}\,, \sqrt{s_2} = (1417^{+4}_{-4} - i\,24^{+7}_{-4})\ \text{MeV}\,, \quad \text{(Fit I)}$$
$$\sqrt{s_1} = (1388^{+9}_{-9} - i\,114^{+24}_{-25})\ \text{MeV}\,, \sqrt{s_2} = (1421^{+3}_{-2} - i\,19^{+8}_{-5})\ \text{MeV}\,. \quad \text{(Fit II)}$$

We can conclude that the narrower resonance has a pole position $\sqrt{s_2}$ which is quite model-independent. While the broader one at $\sqrt{s_1}$ is more sensitive to the specific model details. Furthermore, in Table 1 we give the couplings of the $\Lambda(1405)$ and $\Lambda(1670)$ resonances evaluated as the residues of the amplitudes at the corresponding pole positions.[11]

Though small discrepancies are observed for $\Lambda(1405)$, roughly we can still conclude that the baryon resonances with $I = 0$ resulting from the two fits are rather compatible. On the contrary, no definite conclusions for the isovector resonances from the two fits can be made. This is because we clearly observe different Σ-types of baryon resonances in the two fits.[11] In Fit II, two narrow Σ resonances around the $\bar{K}N$ threshold are found, while they totally disappear in Fit I. The former case is closer to the leading order

Table 1. The couplings of $|\beta_i|_{(I)}$ for the $\Lambda(1405)$ and $\Lambda(1670)$ from both fits. The numbers are expressed in units of GeV. The isospin subscripts are explicitly shown when two or more isospin channels exist.

Pole	$|\beta_{\pi\Lambda}|$	$|\beta_{\pi\Sigma}|0$	$|\beta_{\pi\Sigma}|1$	$|\beta_{\pi\Sigma}|2$	$|\beta_{\bar{K}N}|0$	$|\beta_{\bar{K}N}|1$	$|\beta_{\eta\Lambda}|$	$|\beta_{\eta\Sigma}|$	$|\beta_{K\Xi}|0$	$|\beta_{K\Xi}|1$
$\Lambda(1405)$ Fit I: $\sqrt{s}_1$	$0.0^{+0.0}_{-0.0}$	$8.8^{+0.9}_{-0.4}$	$0.0^{+0.0}_{-0.0}$	$0.0^{+0.0}_{-0.0}$	$7.7^{+1.3}_{-0.7}$	$0.0^{+0.1}_{-0.0}$	$1.4^{+0.4}_{-0.3}$	$0.0^{+0.1}_{-0.0}$	$2.1^{+0.8}_{-0.7}$	$0.0^{+0.0}_{-0.0}$
$\sqrt{s}_2$	$0.1^{+0.0}_{-0.0}$	$5.0^{+1.5}_{-0.8}$	$0.1^{+0.0}_{-0.0}$	$0.0^{+0.0}_{-0.0}$	$7.7^{+1.2}_{-0.6}$	$0.1^{+0.0}_{-0.0}$	$1.4^{+0.4}_{-0.3}$	$0.1^{+0.0}_{-0.0}$	$1.5^{+0.7}_{-0.5}$	$0.1^{+0.0}_{-0.0}$
$\Lambda(1405)$ Fit II: $\sqrt{s}_1$	$0.0^{+0.0}_{-0.0}$	$8.2^{+0.8}_{-0.5}$	$0.0^{+0.0}_{-0.0}$	$0.0^{+0.0}_{-0.0}$	$6.1^{+1.1}_{-0.6}$	$0.1^{+0.0}_{-0.0}$	$2.2^{+0.6}_{-0.3}$	$0.0^{+0.0}_{-0.0}$	$1.9^{+0.2}_{-0.1}$	$0.1^{+0.0}_{-0.0}$
$\sqrt{s}_2$	$0.2^{+0.1}_{-0.1}$	$4.2^{+1.5}_{-0.9}$	$0.2^{+0.0}_{-0.0}$	$0.0^{+0.0}_{-0.0}$	$6.2^{+1.2}_{-0.5}$	$0.3^{+0.1}_{-0.1}$	$2.8^{+0.5}_{-0.3}$	$0.4^{+0.2}_{-0.1}$	$0.7^{+0.4}_{-0.3}$	$0.4^{+0.1}_{-0.1}$
$\Lambda(1670)$ Fit I	$0.0^{+0.0}_{-0.0}$	$0.9^{+0.4}_{-0.2}$	$0.0^{+0.0}_{-0.0}$	$0.0^{+0.0}_{-0.0}$	$1.6^{+0.4}_{-0.2}$	$0.0^{+0.0}_{-0.0}$	$1.7^{+0.5}_{-0.3}$	$0.0^{+0.0}_{-0.0}$	$11.1^{+0.3}_{-0.3}$	$0.1^{+0.0}_{-0.0}$
$\Lambda(1670)$ Fit II	$0.0^{+0.0}_{-0.0}$	$0.8^{+0.1}_{-0.1}$	$0.1^{+0.0}_{-0.0}$	$0.0^{+0.0}_{-0.0}$	$1.6^{+0.4}_{-0.4}$	$0.1^{+0.0}_{-0.0}$	$1.8^{+0.2}_{-0.2}$	$0.1^{+0.0}_{-0.0}$	$10.5^{+0.2}_{-0.2}$	$0.1^{+0.0}_{-0.0}$

study of Ref. 1, so that it is preferred from the point of view of stability of the unitarized chiral perturbation theory expansion. The new experimental inputs from the $\pi\Sigma$ event distributions measured in the photoproduction[6] may provide a useful constraint to discriminate the two solutions.

This work is partially funded by the grants National Natural Science Foundation of China (NSFC) under contract Nos. 11105038 and 11075044, Natural Science Foundation of Hebei Province with contract No. A2011205093 and Doctor Foundation of Hebei Normal University with contract No. L2010B04. We would like to thank partial funding to the (Spanish) projects FPA2010-17806 and Fundación Séneca 11871/PI/09.

References

1. J. A. Oller and U. G. Meissner, Phys. Lett. B **500** (2001) 263.
2. J. A. Oller, Eur. Phys. J. A **28** (2006) 63.
3. Y. Ikeda, T. Hyodo and W. Weise, Phys. Lett. B **706** (2011) 63.
4. Y. Ikeda, T. Hyodo and W. Weise, Nucl. Phys. A **881** (2012) 98.
5. M. Mai and U. -G. Meissner, Nucl. Phys. A **900** (2013) 51.
6. K. Moriya *et al.* [CLAS Collaboration], Phys. Rev. C **87** (2013) 035206.
7. H. Y. Lu *et al.* [CLAS Collaboration], Phys. Rev. C **88** (2013) 045202.
8. M. Bazzi, *et al.*, Phys. Lett. B **704** (2011) 113.
9. M. Iwasaki *et al.*, Phys. Rev. Lett. **78** (1997) 3067; T. M. Ito, *et al.*, Phys. Rev. C **58** (1998) 2366.
10. G. Beer *et al.*, DEAR Collaboration, Phys. Rev. Lett. **94** (2005) 212302.
11. Z. -H. Guo and J. A. Oller, Phys. Rev. C **87** (2013) 035202.
12. U. G. Meissner, U. Raha and A. Rusetsky, Eur. Phys. J. C **35** (2004) 349.

Tensor force and Delta excitation for the structure of light nuclei

K. Horii[*] and H. Toki[†]

*Research Center for Nuclear Physics, Osaka University,
Ibaraki city, Osaka, Japan*
[*] *E-mail: horii@rcnp.osaka-u.ac.jp*
[†] *E-mail: toki@rcnp.osaka-u.ac.jp*

T. Myo

*Osaka Institute of Technology University,
Osaka city, Osaka, Japan
E-mail: myo@ge.oit.ac.jp*

We treat explicitly $\Delta(1232)$ isobar degrees of freedom using bare nucleon-nucleon interaction, which can be the origin of the three-body forces via the pion exchange. We adopt the Argonne delta model potential (AV28) and study the explicit role of Δ in light nuclei. It is surprising that the additional Δ states generate strong tensor correlations from the transitions between N and Δ states, and change various matrix elements from the results with only the nucleon space.

Keywords: Tensor force; delta excitation; nucleon-nucleon interaction; three-body force.

1. Introduction

It is important to understand the structure of nuclear many-body systems by treating important characteristics of nucleon-nucleon(NN) interaction, such as a strong short-range repulsion and a pion-exchange force. It is well known that the nuclear binding energies by using two-body NN interaction such as AV18 potential are less than the experimental values. The shortage of the energies has been handled by means of the effect of three-body interactions phenomenologically.[1] A meson-exchange model for three-nucleon force has been proposed by Fujita and Miyazawa[2]. This model derives the three-nucleon interaction from the second order pion-exchange, where the intermediate nucleon is excited to the P-wave Δ resonance. Hence Δ plays an important role to generates the intermediate-range attraction treated as the three-body force. On the basis of the Fujita-Miyazawa model, most models of three-body force have been developed phenomenologically constructed, where the Δ excitations are renormalized effectively in the nucleon space. In this way we would like to introduce explicitly $\Delta(1232)$ degrees of freedom as the origin of the three-body force in nuclei. It is important to develop a model with the explicit treatment of Δ for many-body framework.

To deal with the two-body interaction including explicitly Δ degrees of freedom, we extend the model space of NN into NN, $N\Delta$ and $\Delta\Delta$ channels. Similar approach has been implemented by the Hannover and Los Alamos groups, and they have investigated the effects of Δ and three-body force in the three-body system.[4,5] In the present study, we would like to examine the roles of tensor force caused by a pion-exchange via Δ excitation. The importances of tensor force are not only giving large attraction in energy but also inducing high momentum components of nucleons in nuclei. Hence it would be possible to generate a large tensor force with the excitation of the nucleon into the Δ state.

2. Effects of Delta for Two-body System

In order to understand the roles of Δ in nuclei, we start with deuteron including explicit Δ degrees of freedom. The wave function of deuteron consisting of NN and $\Delta\Delta$ states with the isospin $T = 0$ channel is given as,

$$|\Psi\rangle = |\Psi_{NN}\rangle + |\Psi_{\Delta\Delta}\rangle \tag{1}$$
$$|\Psi_{NN}\rangle = |^3S_1\rangle + |^3D_1\rangle$$
$$|\Psi_{\Delta\Delta}\rangle = |^3S_1\rangle + |^3D_1\rangle + |^7D_1\rangle + |^7G_1\rangle .$$

We present the energy and various Hamiltonian matrix elements using AV28 potential,[3] and probabilities of the states in Table 1. We also compare the results with those of AV14 potential.[3] Both results provide the same binding energies of 2.2 MeV, but, the matrix elements are largely different. Especially the tensor force component becomes large with AV28 as -35.3 MeV in comparison with the value of AV14. The enhancement of tensor force also increases the kinetic energy, where the contributions of mass difference between NN and $\Delta\Delta$ is obtained as 3.1 MeV. This small mass difference from the original value of about 600 MeV is due to the small probabilities of 0.5 % for the $\Delta\Delta$ state. However even the small mixing of Δ generates the large tensor matrix elements in deuteron. We find that the additional Δ comprises about -10 MeV of the tensor matrix elements, and the transitions between $^3S_1$ of NN state and $^7D_1$ of $\Delta\Delta$ state play a important roles. The central force components provide the repulsive effect to cancel out the large attraction of tensor force, which is an opposite sign to the results of AV14 potential.

Table 1. Calculations of deuteron with AV14 and AV28. The upper part shows various energy components in unit of MeV, and the lower one is the probabilities of the wave function.

^{2}H	Energy	Kinetic	Central	Tensor	LS	L^2	$(LS)^2$
AV14	−2.2	19.1	−1.9	−18.8	0.4	3.1	−4.0
AV28	−2.2	23.9	8.8	−35.3	0.8	3.5	−4.1

P [%]	^{3}S(NN)	^{3}D(NN)	^{3}S($\Delta\Delta$)	^{3}S($\Delta\Delta$)	^{7}D($\Delta\Delta$)	^{7}G($\Delta\Delta$)
AV14	93.9	6.1				
AV28	93.3	6.2	0.04	0.02	0.42	0.04

In order to examine the properties of $N\Delta$ states with the $T = 1$ channel, we discuss another two-body system of the singlet-even (^1E) channel. The wave function of ^1E channel with $J=0$ consists of NN, $N\Delta$ and $\Delta\Delta$ states,

$$|\Psi\rangle = |\Psi_{NN}\rangle + |\Psi_{N\Delta}\rangle + |\Psi_{\Delta\Delta}\rangle \qquad (2)$$
$$|\Psi_{NN}\rangle = |^1S_0\rangle$$
$$|\Psi_{N\Delta}\rangle = |^5D_0\rangle$$
$$|\Psi_{\Delta\Delta}\rangle = |^1S_0\rangle + |^5D_0\rangle \ .$$

Since the ^1E channel is unbound, we analyze this state using the Hamiltonian with radius constraints. Then we restrain the two-body system with a similar radius of deuteron. As a result we obtain the large tensor energy of -14 MeV, where total energy is 3 MeV with kinetic energy as 14 MeV. The NN wave function consisting of only S-wave does not make the tensor correlations by itself, hence this large tensor components come from the transitions between N and Δ. Particularly we find the couplings between $^1S_0$ of NN state and $^5D_0$ of $N\Delta$ state are most significant.

3. Triton Including Delta Degrees of Freedom

We would like to discuss the structure of triton including Δ degrees of freedom. Our model space allows up to double Δ excitation, and the wave function is described as,

$$|\Psi\rangle = |\Psi_{NNN}\rangle + |\Psi_{N\Delta N}\rangle + |\Psi_{\Delta\Delta N}\rangle \ . \qquad (3)$$

For $N\Delta N$ and $\Delta\Delta N$ states, we take into account the Δ excitation of all nucleons due to the antisymmetrization of the particle. In this study, we work out variational calculation to obtain the triton state by using Stocathtic Variational Method.[6] There are many configurations for spin and isospin channels due to the additional Δ state. Hence it is important to make the energy convergence faster by choosing the efficient configurations which bring the large tensor correlations, based on the concept of Tensor Optimized Shell Model.[7,8]

Table 2 shows the results of triton including Δ degrees of freedom using AV28 potential and are compared with those of AV14 potential. In this study, we have not finished the calculations of LS, L^2, $(LS)^2$ components yet, hence the binding energies are smaller than the experimental value of -8.4 MeV. Both potentials provide similar binding energies, but the Hamiltonian components are different significantly. In the results with AV28 potential, the tensor component surprisingly increases, which is comparable to the large repulsion of the kinetic and central energies. About -40 MeV of the tensor components come from the additional Δ in the $N\Delta N$ and $\Delta\Delta N$ states. It is also interesting that this tensor component is caused mainly by the transition between NNN states and $N\Delta N, \Delta\Delta N$ states. Not only in deuteron but also triton cases, we find that the transitions from nucleon to Δ make strong tensor correlations and contribute considerably to the binding energy.

Table 2. Calculations of triton with AV14 and AV28. The table shows the binding energies and various energy components in unit of MeV.

^{3}H	Energy	Kinetic	Central	Tensor
AV14	-6.7	42.7	-17.3	-32.1
AV28	-7.1	58.6	22.4	-88.1

References

1. S.C. Pieper and R.B. Wiringa, *Annu. Rev. Nucl. Part. Sci.* **51**, 53 (2001).
2. J. Fujita and H. Miyazawa, *Prog. Theo. Phys.* **17**, 360 (1957).
3. R.B. Wiringa, R.A. Smith, and T.L. Ainsworth, *Phys. Rev. C* **29**, 1207 (1984).
4. Ch. Hajduk, P.U. Sauer, and W. Strueve, *Nucl. Phys. A* **405**, 581 (1983).
5. A. Picklesminer, R.A. Rice, and R. Brandenburg, *Phys. Rev. Lett* **68**, 1484 (1992).
6. Y. Suzuki *et al*, *Few Body Syst.* **42**, 33 (2008).
7. T. Myo, H. Toki, K. Ikeda, *Prog. Theo. Phys.* **121**, 511 (2009).
8. K. Horii, T. Myo, H. Toki, K. IKeda, *Prog. Theo. Phys.* **127**, 1109 (2011).

Compositeness of hadron resonances in chiral dynamics*

Tetsuo Hyodo

Yukawa Institute for Theoretical Physics, Kyoto University, Kyoto 606-8502, Japan
E-mail: hyodo@yukawa.kyoto-u.ac.jp

Recent experimental observations of many unconventional hadronic states stimulate an interest in the structure of hadrons. While various internal configurations have been proposed, it is a subtle problem to identify the structure of hadron resonances in a model independent manner. Here we discuss the composite/elementary nature of hadrons using the field renormalization constant Z. In particular, we show that the magnitude of the effective range parameter r_e is related to the structure of s-wave near-threshold resonances.

Keywords: Hadron structure; compositeness; resonances; effective range.

1. Introduction

Over the past few years, there have been remarkable developments in the hadron spectroscopy.[1] High-energy inclusive experiments produce evidences of many new exotic hadrons in the heavy quark sector,[2] and the low-energy exclusive experiments provide precise data of the production cross sections and the mass spectra.[3] These results indicate that the spectrum of hadrons is much richer than expected in conventional constituent quark models, and the existence of exotic configurations, such as multiquarks and hadronic molecules, is more and more convinced. The study of exotic forms of hadrons is one of the central issues in hadron physics.

Traditionally, the hadron structure has been studied by model calculations with various configurations. It has been however realized that these studies cannot give the final conclusion, because of the model-dependent nature of the argument. If available, a model-independent classification is superior, as it will ultimately allow us to verify the hadron structure directly by experimental observables and/or lattice QCD simulation. One promising approach is to focus on the compositeness of hadrons suggested in Ref. 4. This method provides a model-independent measure of the structure of weakly bound states, which can be related to experimental observables.

Here we note that the the stable hadrons are only the small fraction of the observed states, and a vast amount of hadrons are unstable against the decay via strong interaction.[1] In particular, the most of candidates of exotic hadrons are resonances with finite decay width. Thus, our task is to clarify the structure of hadron resonances on firm ground. To this end, we generalize the idea of compositeness to resonances, and discuss their structure in relation with experimental observables.[5–7]

2. Compositeness of Bound States

We first introduce the compositeness of stable bound states based on Ref. 4. We focus on three aspects in this approach; the meaning and the model dependence of the field renormalization constant Z (Subsec. 2.1), the model-independent relation with threshold parameters in the weak-binding limit (Subsec. 2.2), and the interpretation of the effective range parameter r_e in the weak-binding limit (Subsec. 2.3). A detailed account on the formulation of this approach can be found in the recent review article.[7]

2.1. Definition of the field renormalization constant

For illustration, we consider the scattering system described by a Hamiltonian H whose eigenstates include one bound state $|B\rangle$:

$$H|B\rangle = -B|B\rangle, \tag{1}$$

where B is the binding energy. To define the field renormalization constant, we first decompose the Hamiltonian H into two parts, the free part H_0 and the interaction V, as

$$H = H_0 + V. \tag{2}$$

The eigenstates of the free Hamiltonian H_0 consist of the two-body scattering states labeled by the momentum $|p\rangle$. The eigenvalue of $|p\rangle$ is given by $E_p = p^2/2\mu$ where μ is the reduced mass of the two-body system. We also assume that there is one discrete energy level in addition to the two-body continuum, which we call a bare bound state $|B_0\rangle$. These eigenstates of the bare Hamiltonian H_0 form a complete set

$$1 = |B_0\rangle\langle B_0| + \int d\boldsymbol{p}|\boldsymbol{p}\rangle\langle\boldsymbol{p}|, \tag{3}$$

which ensures that the model space is completely spanned by $|B_0\rangle$ and $|\boldsymbol{p}\rangle$.

An important feature of the bound state is that the wavefunction $|B\rangle$ can be normalized as

$$\langle B|B\rangle = 1. \tag{4}$$

Because of this normalization, we can decompose the unity into two parts

$$1 = Z + X, \tag{5}$$

$$Z \equiv \langle B|B_0\rangle\langle B_0|B\rangle = |\langle B_0|B\rangle|^2, \tag{6}$$

$$X \equiv \int d\boldsymbol{p}\langle B|\boldsymbol{p}\rangle\langle\boldsymbol{p}|B\rangle = \int d\boldsymbol{p}|\langle\boldsymbol{p}|B\rangle|^2, \tag{7}$$

where Z is the renormalization constant of the field $|B_0\rangle$. Because Z (X) measures the overlap of the physical bound state $|B\rangle$ with the bare bound state $|B_0\rangle$ (the scattering states $|\boldsymbol{p}\rangle$), we interpret it as elementariness (compositeness) of the bound state $|B\rangle$. We note that Z and X are given by the modulus squared of complex numbers,[a] so that they are real and non-negative. This fact, together with Eq. (5), provides the normalization

$$0 \leq Z \leq 1, \quad 0 \leq X \leq 1. \tag{8}$$

Before proceeding, we make three remarks.

(i) The bare state $|B_0\rangle$ is a discrete energy level in the free Hamiltonian H_0, but one may wonder where does it come from. Possible origin of $|B_0\rangle$ may be a bound state of the channels which are integrated out, or a state formed by some microscopic interactions at more fundamental level.[7] In any case, the dynamical origin of $|B_0\rangle$ comes from the outside of the present model space of $|\boldsymbol{p}\rangle$. In this sense, we can regard $|B_0\rangle$ as the Castillejo–Dalitz–Dyson (CDD) pole contribution.[8]

(ii) The probabilistic interpretation of Z and X is guaranteed by the normalization given in Eqs. (5) and (8). We emphasize that the normalization of the state vector (4) is essential, and this is no longer the case for resonances as discussed below.

(iii) In general, Z and X are model dependent. For a given Hamiltonian H, the decomposition (2) is not unique, and Z and X depends on the choice of this decomposition. For instance, Eq. (7) can be rewritten as

$$X = 1 - Z = \int d\boldsymbol{p}\frac{|\langle\boldsymbol{p}|V|B\rangle|^2}{(E_p + B)^2}, \tag{9}$$

which explicitly depends on the potential V.

---

[a] $\langle B_0|B\rangle$ and $\langle\boldsymbol{p}|B\rangle$ are in general complex.

2.2. Weak-binding limit

We have seen that the elementariness Z and compositeness X are well-defined once the model space is specified. Interestingly, however, Ref. 4 found that the model dependence does not appear for weakly bound states. We mean by weak-binding that the binding energy B is much smaller than the typical energy scale of the interaction E_{typ}:

$$B \ll E_{\text{typ}}. \tag{10}$$

The magnitude of E_{typ} depends on the underlying physics of the two-body system. For instance, the long range part of the nuclear force is governed by the pion exchange which provides $E_{\text{typ}} = m_\pi^2/M_N \sim 20$ MeV. Thus, we may regard the deuteron with $B \sim 2$ MeV as a weakly bound state.

When the condition (10) is satisfied, Z and X are model-independently related to the threshold parameters of the two-body scattering:[4,7]

$$a = \frac{2X}{X+1}R + \mathcal{O}(R_{\text{typ}}) = \frac{2(1-Z)}{2-Z}R + \mathcal{O}(R_{\text{typ}}), \tag{11}$$

$$r_e = \frac{X-1}{X}R + \mathcal{O}(R_{\text{typ}}) = \frac{-Z}{1-Z}R + \mathcal{O}(R_{\text{typ}}), \tag{12}$$

where we define $R \equiv 1/\sqrt{2\mu B}$ and $R_{\text{typ}} \equiv 1/\sqrt{2\mu E_{\text{typ}}}$ which represents the typical length scale of the interaction. The scattering length a and the effective range r_e are defined in the s-wave scattering amplitude as

$$f(p) = \left[-\frac{1}{a} - ip + \frac{r_e}{2}p^2 \right]^{-1}. \tag{13}$$

It is remarkable that the all the information of the potential V is included in the threshold parameters. Because a and r_e are measurable in experiments, we can determine X and Z from experimental observables.

2.3. Interpretation of threshold parameters

To interpret the results in the weak-binding limit, we consider the behavior of the threshold parameters for a given structure of the bound state. The dominance of the elementary (composite) component is characterized by $Z \sim 1$ and $X \sim 0$ ($Z \sim 0$ and $X \sim 1$). Hence, Eqs. (11) and (12) lead to

$$\begin{cases} a \sim R_{\text{typ}} \ll -r_e & \text{(elementary dominance)}, \\ a \sim R \gg r_e \sim R_{\text{typ}} & \text{(composite dominance)}. \end{cases} \tag{14}$$

This indicates that the structure of the bound state can be judged through the comparison of a and r_e.

Naively, the existence of a bound state close to the threshold leads to a large scattering length a. If we consider the bound state formed by a potential having only the attractive component, we can adjust the binding energy by modifying the depth of the potential, without changing the effective range which is related to the spatial extent of the potential. By decreasing B, the scattering length is enhanced and we always obtain the composite dominance. This is consistent with the expectation that the bound state produced by the potential scattering has a composite nature.

A characteristic feature of the elementary dominance is the large and negative effective range r_e. In particular, the magnitude of r_e should be much larger than the typical length scale of the interaction R_{typ}. This can be achieved by the strong energy- and/or momentum-dependence of the interaction.[9,10] For example, if there is a bare bound state $|B_0\rangle$ near the threshold, the two-body interaction contains a pole term which produces a strong energy dependence. We therefore conjecture that a large negative effective range r_e is a consequence of the existence of substantial elementary component, which originates in the bare state $|B_0\rangle$.

3. Generalization to Resonances

As mentioned, we need to generalize this framework to the case of resonances. It is however noticed that X and Z of resonances are given by complex numbers, and the interpretation of these quantities is not clear. In fact, this is related to an inherent nature of resonances.[7] In contrast to the bound state case, the state vector of a resonance $|R\rangle$ is not normalizable. A kind of normalization is achieved by introducing the antiresonance state $|\tilde{R}\rangle = |R^*\rangle$ which is the solution of the same eigenvalue equation with a different boundary condition.[11-13] The normalization is then given by

$$\langle \tilde{R}|R\rangle = 1. \tag{15}$$

Following the same discussion with Subsec. 2.1, we obtain

$$1 = Z + X, \tag{16}$$

$$Z \equiv \langle \tilde{R}|B_0\rangle\langle B_0|R\rangle = \langle B_0|R\rangle^2, \tag{17}$$

$$X \equiv \int d\boldsymbol{p}\langle \tilde{R}|\boldsymbol{p}\rangle\langle \boldsymbol{p}|R\rangle = \int d\boldsymbol{p}\langle \boldsymbol{p}|R\rangle^2. \tag{18}$$

In contrast to Eqs. (6) and (7), Z and X are *not* the modulus squared, but the complex number squared. This leads to the complex Z and X, and the probabilistic interpretation of these quantities is no longer possible in the strict sense.

3.1. *Applications to hadron resonances*

The generalization of the compositeness approach to resonances has been formulated, for instance, by the integration of the spectral density.[14,15] Another method is to evaluate the field renormalization constant Z on top of the resonance pole in the complex energy plane.[5] For this purpose, we express Eq. (9) in the field theoretical model with Yukawa interaction as[5]

$$X = 1 - Z = -g^2 \frac{dG(W)}{dW}\bigg|_{W \to M_B}, \qquad (19)$$

where W is the total energy of the two-body scattering and M_B is the mass of the bound state, and $G(W)$ is the two-body loop function. The coupling strength g^2 can be evaluated by the residue of the pole of the scattering amplitude $T(W)$ as $g^2 = \lim_{W \to M_B}(W - M_B)T(W)$. Note that the form of the loop function depends on the regularization scheme as well as the choice of the interaction Lagrangian in the Yukawa model. This ambiguity corresponds to the model dependence (V dependence) in Eq. (9).

Generalization of Eq. (19) to a resonance at $W = z_R \in \mathbb{C}$ is given by

$$X = 1 - Z = -g_{\mathrm{II}}^2 \frac{dG_{\mathrm{II}}(W)}{dW}\bigg|_{W \to z_R}, \qquad (20)$$

where the loop function $G_{\mathrm{II}}(W)$ and the coupling strength $g_{\mathrm{II}}^2 = \lim_{W \to z_R}(W - z_R)T(W)$ should be evaluated in the second Riemann sheet of the complex energy plane. In general, Eq. (19) gives a real and nonnegative number for a real M_B below the threshold.[5] On the other hand, X in Eq. (20) is in general complex, which reflects the property of the general expression (18).

It is worth noting that the relation (16) holds even for complex X and Z, as a consequence of the normalization (15). In fact, the generalized Ward identity of the one-photon-attached amplitude gives[16]

$$1 = -g_{\mathrm{II}}^2 \left(\frac{dG_{\mathrm{II}}(W)}{dW} + G_{\mathrm{II}}(W)\frac{dV(W)}{dW}G_{\mathrm{II}}(W) \right)\bigg|_{W \to z_R}, \qquad (21)$$

where $V(W)$ is the interaction kernel. The first term of Eq. (21) corresponds to X and the second term to Z. It is also seen from this expression that the strong energy dependence of $V(W)$ leads to a large Z.

In recent works,[6,17–20] the field renormalization constant Z has been calculated for various hadron resonances as compiled in Table 1. In some cases the imaginary part of Z is small, and its real part or the absolute value may reflect the elementariness of resonances. We should however keep in

Table 1. Field renormalization constant Z of the hadron resonances evaluated on the resonance pole from Ref. 7. The momentum cutoff q_{max} is chosen to be 1 GeV for the $\rho(770)$ and $K^*(892)$ mesons,[17,19] 0.5 GeV for the $\Delta(1232)$ baryon, and 0.45 GeV for the $\Sigma(1385)$, $\Xi(1535)$, Ω baryons.[20]

Baryons	Z	$|Z|$	Mesons	Z	$|Z|$
$\Lambda(1405)$ higher pole (Ref. 18)	$0.00 + 0.09i$	0.09	$f_0(500)$ or σ (Ref. 18)	$1.17 - 0.34i$	1.22
$\Lambda(1405)$ lower pole (Ref. 18)	$0.86 - 0.40i$	0.95	$f_0(980)$ (Ref. 18)	$0.25 + 0.10i$	0.27
$\Delta(1232)$ (Ref. 20)	$0.43 + 0.29i$	0.52	$a_0(980)$ (Ref. 18)	$0.68 + 0.18i$	0.70
$\Sigma(1385)$ (Ref. 20)	$0.74 + 0.19i$	0.77	$\rho(770)$ (Ref. 17)	$0.87 + 0.21i$	0.89
$\Xi(1535)$ (Ref. 20)	$0.89 + 0.99i$	1.33	$K^*(892)$ (Ref. 19)	$0.88 + 0.13i$	0.89
Ω (Ref. 20)	0.74	0.74			
$\Lambda_c(2595)$ (Ref. 6)	$1.00 - 0.61i$	1.17			

mind that, strictly speaking, the interpretation of Z as elementariness is not on firm ground, because of the complex nature of Z for resonances.

3.2. Near-threshold resonances

In the case of bound states, taking the weak-binding limit has some nice features. It is therefore interesting to consider the corresponding case for resonances, in which the resonance pole appears close to the threshold.[6] As in the case of the weakly bound state, the effective range expansion is valid to describe the near-threshold resonance. The pole positions of the amplitude (13) are easily obtained as

$$p^\pm = \frac{i}{r_e} \pm \frac{1}{r_e}\sqrt{\frac{2r_e}{a} - 1}, \tag{22}$$

whose properties have been classified in Ref. 21. The resonance solution is obtained for $r_e < a < 0$, in which the solutions have the property $p^+ = -(p^-)^*$. In other words, the real and imaginary parts of the resonance pole determine the scattering length and the effective range.

Let us take an example of the $\Lambda_c(2595)$ resonance which appears slightly above the $\pi\Sigma_c$ threshold. The central values of the excitation energy and the decay width are $E = 0.67$ MeV and $\Gamma = 2.59$ MeV.[1] Using $p\pm = [2\mu(E \pm i\Gamma/2)]^{1/2}$, we obtain the threshold parameters as

$$a = -\frac{p^+ + p^-}{ip^+ p^-} = -10.5 \text{ fm}, \quad r_e = \frac{2i}{p^+ + p^-} = -19.5 \text{ fm}. \tag{23}$$

According to the discussion in Subsec. 2.3, the large negative effective range indicates that the $\Lambda_c(2595)$ resonance is not likely a $\pi\Sigma_c$ molecule. Even though the field renormalization constant Z becomes complex (see Table 1), the threshold parameters are obtained as real numbers, which provides an interpretation of the result.

4. Conclusions

We have discussed the structure of hadron resonances from the viewpoint of the compositeness. It is shown that the model-independent discussion on the hadron structure is possible for bound states in the weak-binding limit. Although the field renormalization constant becomes complex for resonances, the consideration of the near-threshold resonances enables us to interpret the structure of resonances from the magnitude of the effective range parameter.

Acknowledgments

The author thanks Lisheng Geng, Bingsong Zou, and the organizers of "The Seventh International Symposium on Chiral Symmetry in Hadrons and Nuclei", for their kind hospitality at Beijing. This work was partially supported by JSPS KAKENHI Grant Numbers 24105702 and 24740152, and by the Yukawa International Program for Quark-Hadron Sciences (YIPQS).

References

1. J. Beringer *et al.*, *Phys. Rev. D* **86**, p. 010001 (2012).
2. N. Brambilla *et al.*, *Eur. Phys. J. C* **71**, p. 1534 (2011).
3. T. Hyodo and D. Jido, *Prog. Part. Nucl. Phys.* **67**, 55 (2012).
4. S. Weinberg, *Phys. Rev.* **137**, B672 (1965).
5. T. Hyodo, D. Jido and A. Hosaka, *Phys. Rev. C* **85**, p. 015201 (2012).
6. T. Hyodo, *Phys. Rev. Lett.* **111**, p. 132002 (2013).
7. T. Hyodo, *Int. J. Mod. Phys. A* **28**, p. 1330045 (2013).
8. L. Castillejo, R. H. Dalitz and F. J. Dyson, *Phys. Rev.* **101**, 453 (1956).
9. D. R. Phillips, S. R. Beane and T. D. Cohen, *Annals Phys.* **263**, 255 (1998).
10. E. Braaten, M. Kusunoki and D. Zhang, *Annals Phys.* **323**, 1770 (2008).
11. N. Hokkyo, *Prog. Theor. Phys.* **33**, 1116 (1965).
12. T. Berggren, *Nucl. Phys. A* **109**, 265 (1968).
13. A. Bohm, *J. Math. Phys.* **22**, p. 2813 (1981).
14. V. Baru, J. Haidenbauer, C. Hanhart, Y. Kalashnikova and A. E. Kudryavtsev, *Phys. Lett. B* **586**, 53 (2004).
15. C. Hanhart, Y. S. Kalashnikova and A. V. Nefediev, *Eur. Phys. J. A* **47**, 101 (2011).
16. T. Sekihara, T. Hyodo and D. Jido, *Phys. Rev. C* **83**, p. 055202 (2011).
17. F. Aceti and E. Oset, *Phys. Rev. D* **86**, p. 014012 (2012).
18. T. Sekihara and T. Hyodo, *Phys. Rev. C* **87**, p. 045202 (2013).
19. C. Xiao, F. Aceti and M. Bayar, *Eur. Phys. J. A* **49**, p. 22 (2013).
20. F. Aceti, L. Dai, L. Geng, E. Oset and Y. Zhang (2013).
21. Y. Ikeda, T. Hyodo, D. Jido, H. Kamano, T. Sato and K. Yazaki, *Prog. Theor. Phys.* **125**, 1205 (2011).

Thermodynamics of hadrons using the Gaussian functional method in the linear sigma model

Shotaro Imai

Department of Physics, Graduate School of Science, Kyoto University,
Kitashirakawa-oiwake, Sakyo, Kyoto 606-8502, Japan
E-mail: imai@ruby.scphys.kyoto-u.ac.jp

Hua-Xin Chen[1], Hiroshi Toki[2] and Li-Shen Geng[1]

[1] *School of Physics and Nuclear Energy Engineering and International Research Center*
for Nuclei and Particles in the Cosmos, Beihang University,
Beihang 10091, China
[2] *Reseach Center for Nuclear Physics, Osaka University,*
Ibaraki, Osaka 567-0047, Japan

We investigate thermodynamics of hadrons using the Gaussian functional method (GFM) at finite temperature. Since the interaction among mesons is very large, we take into account fluctuations of mesons around their mean field values using the GFM. We obtain the ground state energy by solving the Schrödinger equation. The meson masses are obtained using the energy minimization condition. The resulting mass of pion is not zero even in the spontaneous chiral symmetry broken phase due to the non-perturbative effect. We consider then the bound state of mesons using the Bethe-Salpeter equation and show that the Nambu-Goldstone theorem is recovered. We investigate further the behavior of the meson masses and the mean filed value as functions of temperature for the cases of chiral limit and explicit chiral symmetry breaking.

Keywords: Linear sigma model; Nambu-Goldstone theorem; Gaussian functional method; Bethe-Salpeter equation; finite temperature.

1. Introduction

The linear sigma model is often used to discuss chiral symmetry breaking in the mean field approximation. The interaction among mesons, however, is very large. The fluctuations of mesons around their mean field values should be included in a beyond mean field theory. We use the Gaussian functional (GF) method[1] as a non-perturbative many body theory. The ground state energy is obtained by solving the Schrödinger equation with the trial Gaussian wave functional in the GF method. The gap equation for the field vacuum expectation values and the dressed masses are determined by the minimization condition of the energy. The resulting mass of Nambu-Goldstone boson is, however, not zero, even the chiral symmetry is spontaneously broken in the chiral limit. In order to recover the Nambu-Goldstone theorem, we have to solve the Bethe-Salpeter (BS) equation.[2,3]

In this paper we shall study hadron properties at finite temperature using the BS equation after solving the GF equations for the two cases of chiral limit and explicit chiral symmetry breaking.

2. The Linear Sigma Model in the Gaussian Functional Method

We start from the $O(4)$ symmetric linear sigma model

$$\mathcal{L} = \frac{1}{2}(\partial_\mu \phi)^2 + \frac{1}{2}\mu_0 \phi^2 - \frac{\lambda_0}{4}(\phi^2)^2 + \varepsilon\sigma, \tag{1}$$

where the fields are denoted as the column vector $\phi = (\phi_0, \phi_1, \phi_2, \phi_3) = (\sigma, \boldsymbol{\pi})$. The parameters in the Lagrangian are the mass term μ_0 and the coupling constant λ_0. The explicit chiral symmetry breaking term can be expressed as $\varepsilon\sigma$. The Gaussian functional method (GFM) can take into account radiative corrections of meson loops. We use the following Gaussian ground state wave functional ansatz:

$$\Psi[\phi] = \mathcal{N} \exp\left(-\frac{1}{4\hbar}\int d\boldsymbol{x} d\boldsymbol{y}[\phi_i(\boldsymbol{x}) - \langle\phi_i(\boldsymbol{x})\rangle]G_{ij}^{-1}(\boldsymbol{x},\boldsymbol{y})[\phi_j(\boldsymbol{y}) - \langle\phi_j(\boldsymbol{y})\rangle]\right), \tag{2}$$

where $\langle\phi_i\rangle$ is the vacuum expectation value of the i-th field, and we define

$$G_{ij}(\boldsymbol{x},\boldsymbol{y}) = \frac{1}{2}\delta_{ij}\int \frac{d^3\mathbf{k}}{(2\pi)^3}\frac{1}{\sqrt{\mathbf{k}^2 + M_i^2}}e^{i\mathbf{k}\cdot(\boldsymbol{x}-\boldsymbol{y})}, \tag{3}$$

where M_i is the "dressed" mass of the σ meson (denoted by $M_0 = M_\sigma$) and the π meson ($M_{1,2,3} = M_\pi$). The ground state energy is defined as

$$\mathcal{E}(M_i, \langle\phi_i\rangle) = \int \mathcal{D}\phi \Psi^*[\phi]\mathcal{H}[\phi]\Psi[\phi], \tag{4}$$

where the Hamiltonian $\mathcal{H}$ is defined by the Lagrangian (1) through the Legendre transformation. The variational parameters $\langle\phi_i\rangle$ and M_i are determined by the energy minimization condition,

$$\left(\frac{\partial\mathcal{E}(M_i, \langle\phi_i\rangle)}{\partial\langle\phi_i\rangle, M_i}\right)_{\min} = 0, \text{ for } i = 0\ldots3, \tag{5}$$

which leads to

$$\langle\phi_0\rangle = v, \quad \langle\phi_i\rangle = 0 \ \text{for} \ i = 1, 2, 3 \tag{6}$$

$$M_\sigma^2 = \frac{\varepsilon}{v} + 2\lambda_0 v^2, \tag{7}$$

$$M_\pi^2 = \frac{\varepsilon}{v} + 2\lambda_0 \left[I_0(M_\pi) - I_0(M_\sigma) \right], \tag{8}$$

where the loop integration,

$$I_0(M_i) = i \int^\Lambda \frac{d^4k}{(2\pi)^4} \frac{1}{k^2 - M_i^2 + i\epsilon}, \tag{9}$$

and the momentum cutoff Λ have been introduced. Although the pion mass M_π should be zero, when the chiral symmetry is broken in the chiral limit, it is finite due to the non-perturbative loop correction (see Eq. (8)). We cannot identify the "dressed" pion as the NG boson.

3. The Bethe-Salpeter Equation

The Physical mass can be obtained by solving the Bethe-Salpeter equation. The single channel Bethe-Salpeter equation of the σ-π scattering gives the physical pion mass m_π and the coupled-channels of σ-σ and π-π scatterings give the physical σ mass m_σ. We show the result of the physical masses $m_{\sigma,\pi}$ as a function of the dressed pion mass M_π in the chiral limit in Fig. 1(a), and the explicit breaking case in Fig. 1(b). The Nambu-Goldstone theorem is always fulfilled in the chiral limit in any value of the dressed pion mass. Hence, the physical masses should be considered as a bound state of the dressed particles.

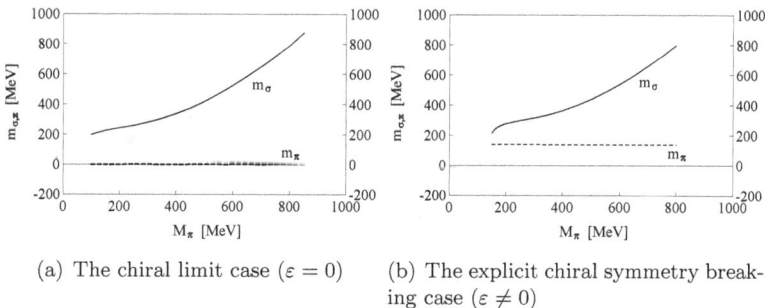

(a) The chiral limit case ($\varepsilon = 0$)   (b) The explicit chiral symmetry breaking case ($\varepsilon \neq 0$)

Fig. 1.   The physical masses m_σ and m_π as a function of the dressed pion mass M_π.

4. Thermodynamics of Hadrons in the GFM+BS Scheme

In this section we discuss the mass spectra at finite temperature. The result is shown in Fig. 2. The physical pion mass m_π stays at zero or 138 MeV until the critical temperature T_c, which means that the chiral symmetry is spontaneously broken. The masses suddenly jump and coincide with each other above T_c, since the chiral symmetry is restored. By solving the Bethe-Salpeter equation, we can see that the physical masses coincide with the dressed masses when the chiral symmetry is restored ($v \to 0$). This behavior suggests that the physical states are bound states of the dressed mesons in the symmetry broken phase, and the physical states become the dressed mesons in the symmetric phase.

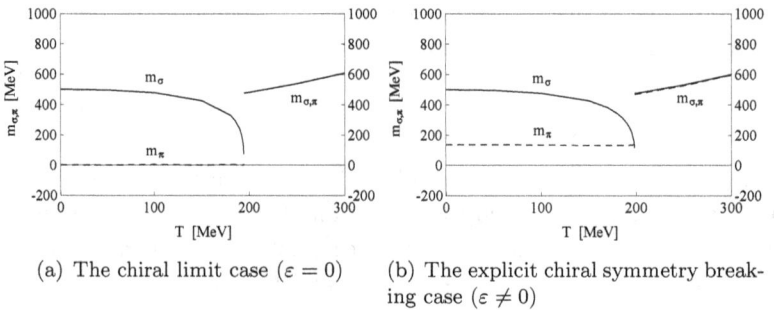

(a) The chiral limit case ($\varepsilon = 0$)   (b) The explicit chiral symmetry breaking case ($\varepsilon \neq 0$)

Fig. 2.   The physical masses m_σ and m_π as a function of temperature.

5. Summary

We have investigated the meson mass spectra at finite temperature using the Gaussian functional method with the Bethe-Salpeter equation in the linear sigma model. The Nambu-Goldstone theorem is fulfilled in this scheme, which is explicitly shown in this study. We discuss the thermodynamical properties of the mesons. The physical states can be considered as the bound states of the dressed mesons in the chiral symmetry broken phase.

References

1. T. Barnes and G. Ghandour, *Phys.Rev.* **D22**, p. 924 (1980).
2. V. Dmitrasinovic, J. Shepard and J. McNeil, *Z.Phys.* **C69**, 359 (1996).
3. I. Nakamura and V. Dmitrasinovic, *Prog.Theor.Phys.* **106**, 1195 (2001).

Studies of hypernuclei with the AMD method

M. Isaka

The Nishina Center for Accelerator-Based Science, RIKEN,
Wako, Saitama 351-0198, Japan
E-mail: masahiro.isaka@riken.jp

Level structure of $^{25}_{\Lambda}$Mg with the Λ particle in p orbit is investigated within the framework of the antisymmetrized molecular dynamics. It is found that there rotational bands appear by adding Λ in p orbit due to the triaxial deformation of the host nucleus ^{24}Mg.

1. Introduction

In this article, we focus on the excited states in $^{25}_{\Lambda}$Mg with the Λ particle in p-orbit which we call p state in the following.

Since a hyperon is unaffected by the nuclear Pauli principle, it can penetrate into nuclear interior and show unique phenomena of hypernuclei. In the last several decades, hypernuclear studies are mainly concentrated in s-shell and p-shell Λ hypernuclei. For example, many authors discussed structure of $^{9}_{\Lambda}$Be hypernuclei.[1,2] In the cluster model picture, $^{9}_{\Lambda}$Be is considered to have a $\alpha+\alpha+\Lambda$ cluster structure, in which the $\alpha+\alpha$ part corresponds to the core nucleus ^{8}Be. By adding Λ to ^{8}Be, three kinds of the rotational bands are predicted to appear. Among them, the 2nd and 3rd lowest bands are generated by the Λ particle in the p orbit, while the lowest corresponds to the ground band of $^{9}_{\Lambda}$Be with the Λ particle in the s orbit. The 2nd lowest is called "super symmetric states"[1] or "genuine hypernuclear states"[2] in which the Λ particle occupies the p orbit parallel to the 2α cluster structure. The Λ in the p orbit perpendicular to the 2α clustering generates the 3rd lowest band, because the binding energy of Λ in the 3rd lowest is smaller than that in the 2nd lowest band. In other words, $^{9}_{\Lambda}$Be is considered to have two p-states bands due to the symmetric 2α structure of ^{8}Be and anisotropy of the p orbit. Considering triaxially deformed Λ hypernuclei, it is expected that three difference bands of the p states appear. ^{24}Mg is a candidate having triaxial deformation, due to the presence of the low-lying 2nd 2^+ state.[3] Therefore our task of the present work is to reveal and predict the level structure of the p-states in $^{25}_\Lambda$Mg.

To investigate it, we use an extended version of the antisymmetric molecular dynamics (AMD) for hypernuclei,[4] and we call it HyperAMD. The detail of the framework of HyperAMD is explained in the next section.

2. Framework

We have performed the HyperAMD calculation basically following our previous work.[5]

The Hamiltonian used in this study is given as,

$$\hat{H} = \hat{T}_N - \hat{T}_g + \hat{V}_{NN} + \hat{V}_{Coul} + \hat{H}_\Lambda, \ \ \hat{H}_\Lambda = \hat{T}_\Lambda + \hat{V}_{\Lambda N}. \tag{1}$$

Here, $\hat{T}_N$, $\hat{T}_\Lambda$ and $\hat{T}_g$ are the kinetic energies of nucleons, Λ and the center-of-mass motion, respectively. We use the Gogny D1S as an effective nucleon-nucleon interaction $\hat{V}_{NN}$. As an effective ΛN interaction $\hat{V}_{\Lambda N}$, we use the central force of the YNG-NF[6] with Fermi momentum $k_F = 1.2$ fm^{-1}.

The intrinsic wave function of a single Λ hypernucleus composed of A nucleons and a Λ hyperon is described by the parity-projected wave function, $\Psi^\pi = \hat{P}^\pi \Psi_{int}$, where $\hat{P}^\pi$ is the parity projector. The intrinsic wave function Ψ_{int} is given as $\Psi_{int} = \Psi_N \otimes \varphi$, where Ψ_N is defined by a Slater determinant of the nucleon single particle wave packets. The Λ single particle wave function φ is represented by a superposition of Gaussian wave packets, as

$$\varphi = \sum_{m=1}^{M} c_m \phi_m (r) , \ \ \phi_m = \prod_{\sigma=x,y,z} \left(\frac{2\nu_\sigma}{\pi} \right)^{\frac{1}{4}} \exp\left\{ -\nu_\sigma \left(r - z_m \right)_\sigma^2 \right\} \chi_m, \tag{2}$$

to describe Λ in p-orbit. Here χ_m is the spin wave functions of Λ.

By using the frictional cooling method, the variation parameters are so determined as to minimize the total energy under two kinds of the constraints. The first is imposed on nuclear quadruple deformation parameters β and γ to obtain the intrinsic wave functions of $^{25}_\Lambda$Mg for given deformation parameters as in our previous work.[5] The other is imposed on the Λ single particle wave function to obtain p states by adding the constraint potential, $V_f = \lambda \sum_f |\varphi_f\rangle\langle\varphi_f|$, which forbids the Λ hyperon occupying the orbit φ_f with sufficiently large value of λ. The actual calculational procedure is as follows. First, we perform the variational calculation with constraint on (β, γ) but without the constraint on the Λ single particle wave function to obtain the lowest energy state of $^{25}_\Lambda$Mg. Then, denoting the Λ single particle

s-orbit obtained by this calculation as φ_1, we perform another variational calculation with the second constraint, $V_f = \lambda|\varphi_1\rangle\langle\varphi_1|$, as well as the constraint on the nuclear deformation. The resulting state is denoted as φ_2, and we complete the calculation for the s-orbit states. We further proceed the calculation by adding the constraint $V_f = \lambda\sum_{f=1,2}|\varphi_f\rangle\langle\varphi_f|$, which produces the third lowest energy state (*i.e.* the lowest p state) φ_3. By continuing this procedure, we obtain two s-orbits (φ_1 and φ_2) and six p-orbits ($\varphi_3, \cdots, \varphi_8$).

The Λ single particle energies as a function of (β,γ) which is defined as,

$$\epsilon_f(\beta,\gamma) = \langle\Psi_f^\pi(\beta,\gamma)|\hat{H}_\Lambda|\Psi_f^\pi(\beta,\gamma)\rangle, \quad f = 3,4,\cdots,8, \tag{3}$$

are two-fold degenerated due to the small ΛN spin-orbit interaction, and hence, we obtain three p states with different spatial distribution which we denote as φ_{p1}, φ_{p2} and φ_{p3} in ascending order of their single particle energies, ϵ_{p1} ($= \epsilon_3 = \epsilon_4$), ϵ_{p2} ($= \epsilon_5 = \epsilon_6$) and ϵ_{p3} ($= \epsilon_7 = \epsilon_8$).

After the variation, we performed the angular momentum projection and the generator coordinate method (GCM) calculation to obtain the excitation spectra of ${}^{25}_{\Lambda}\mathrm{Mg}$.

3. Results and Discussion

The dependence of the Λ single particle energies (ϵ_{p1}, ϵ_{p2} and ϵ_{p3}) on the nuclear quadruple deformation are clearly seen in Fig. 1(b), where they are plotted along the path on the (β,γ) plane shown in Fig. 1(a). In triaxially deformed region ($\beta = 0.48, 0° < \gamma < 60°$), the Λ single particle energies are different among these three states. It is discussed that the Λ single particle energy varies depending on the spatial overlap between the nucleons (N)

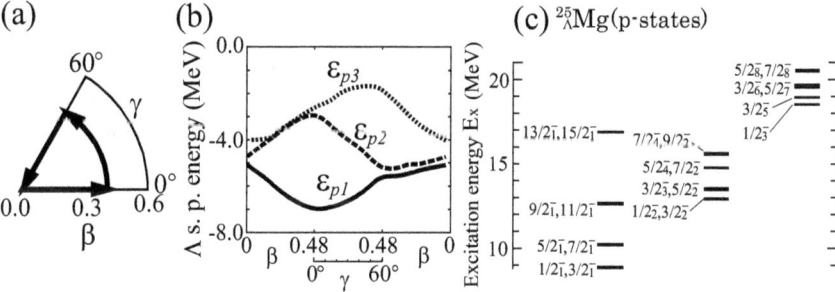

Fig. 1.   (b):Λ single particle energies ϵ_{p1}, ϵ_{p2}, and ϵ_{p3} along the loop shown in (a). This loop starts from and heads back to the origin ($\beta = 0, \gamma = 0°$) via ($\beta = 0.48, \gamma = 0°$) and ($\beta = 0.48, \gamma = 60°$). (c): Excitation spectra of the p states in ${}^{25}_{\Lambda}\mathrm{Mg}$.

and Λ.[4,7] In $^{25}_{\Lambda}$Mg, it is found that the spatial distributions of the Λ are different among the three p states. And thus the spatial overlap between N and Λ is different among the three states and is changed as a function of nuclear quadrupole deformation (β, γ).

To investigate the level structure of the p states, we perform the GCM calculation and the resulting excitation spectrum is shown in Figure 1(c) Three rotational bands are obtained as the p states of $^{25}_{\Lambda}$Mg corresponding to the ground band (GB) of ^{24}Mg. It is found that coupling of φ_{p1}, φ_{p2} and φ_{p3} to the ground band of ^{24}Mg generates the three bands built on the $(1/2_1^-, 3/2_1^-)$, $(1/2_2^-, 3/2_2^-)$ and $(1/2_3^-, 3/2_5^-)$ states, respectively. Therefore, the p states coupled to the ground band of ^{24}Mg split into three bands.

4. Summary

The HyperAMD was applied to $^{25}_{\Lambda}$Mg to investigate the level structure of the p states. After the energy variation, we obtained three p orbit of Λ with different Λ single particle energies. It was found that the difference of the Λ single particle energy comes from the difference of the spatial overlap between N and Λ, since the Λ particle occupies the p orbits with different spatial distributions. By performing the GCM calculation, there different rotational bands of the p states were predicted corresponding to the p orbits with different directions.

Acknowledgments

The author is supported by the Special Postdoctoral Researcher Program of RIKEN.

References

1. R. H. Dalitz and A. Gal, *Phys. Rev. Lett.* **36**, 362 (1976).
2. H. Bandō, M. Seki, and Y. Shōno, *Prog. Theor. Phys.* **66**, 2118 (1981).
3. R. Batchelor, A. Ferguson, H. Gove, and A. Litherland, *Nucl. Phys.* **16**, 38 (1960).
4. M. Isaka, M. Kimura, A. Dote and A. Ohnihi, *Phys. Rev. C* **83**, 044323 (2011).
5. M. Isaka, M. Kimura, A. Dote and A. Ohnihi, *Phys. Rev. C* **85**, 034303 (2012).
6. Y. Yamamoto, T. Motoba, H. Himeno, K. Ikeda and S. Nagata, *Prog. Theor. Phys. Suppl.* **117**, 361 (1994)
7. M. T. Win, K. Hagino, and T. Koike, *Phys. Rev. C* **83**, 014301 (2011).

Light quark mass dependence of the $X(3872)$ in XEFT

M. Jansen*[1,2], H.-W. Hammer†[1,3,4] and Yu Jia‡[2,5]

[1] *Helmholtz-Institut für Strahlen- und Kernphysik,*
Universität Bonn, 53115 Bonn, Germany
[2] *Institute of High Energy Physics, Chinese Academy of Sciences,*
Beijing 100049, China
[3] *Institut für Kernphysik, Technische Universität Darmstadt,*
64289 Darmstadt, Germany
[4] *ExtreMe Matter Institute EMMI,*
GSI Helmholtzzentrum für Schwerionenforschung GmbH,
64291 Darmstadt, Germany
[5] *Theoretical Physics Center for Science Facilities,*
Institute of High Energy Physics, Chinese Academy of Sciences,
Beijing 100049, China
* mjansen@hiskp.uni-bonn.de
† hammer@hiskp.uni-bonn.de
‡ jiay@ihep.ac.cn

Keywords: Heavy quarkonia; chiral symmetries; effective field theory.

The $X(3872)$ is a charmonium-like hadron with a mass close to the $\bar{D}^0 D^{*0}$ threshold. It was first observed in 2003 by the Belle Collaboration[1] and confirmed shortly after by the CDF collaboration.[2] The quantum numbers were recently determined by the LHCb experiment to be $J^{PC} = 1^{++}$.[3] Yet, the nature of the $X(3872)$ is not fully understood. In future, lattice QCD calculations should be able to obtain observables and are expected to contribute to a better understanding of the X.

A first candidate for the $X(3872)$ on the lattice was recently observed by Prelovsek and Leskovec with a mass about 11 ± 7 MeV below the $\bar{D}^0 D^{*0}$ threshold.[4] But the considerably small lattice with a spatial box length of approximately 2 fm and light quark masses at about four times the physical value used in the simulation require finite volume corrections and chiral extrapolations in order to acquire reliable results.

Previous works on the light quark mass dependence of observables for the $X(3872)$ include the use of a unitarized heavy-hadron chiral perturbation theory[5] and numerical studies of non-relativistic Faddeev-type three-body equations,[6] both assuming that the $X(3872)$ is a hadronic molecule of $D^{(*)}$ mesons. While for the former, the authors claim that the X is generated by pion exchange alone and no bound state at high quark masses exists, the latter work finds that observables for the X are sensitive to quark mass dependent contact interactions and the existence of a bound state depends upon the corresponding coupling constants.

We investigated the light quark mass dependence of the $X(3872)$ to next-to-leading order (NLO) using an effective field theory for the $X(3872)$, called XEFT, which was developed by Fleming, Kusunoki, Mehen and van Kolck in 2007.[7] In XEFT, the $X(3872)$ is interpreted as a loosely-bound S-wave hadronic molecule of $D^{(*)}$ mesons, which is supported by the quantum numbers and the proximity of the hadron's mass to the $\bar{D}^0 D^{*0}$ threshold.[8-13] In this picture, the binding energy is given as the difference of the sum of the masses of the $D^{(*)}$ mesons and the mass of the $X(3872)$. Using the latest values for the masses from the review of particle properties,[14] one obtains $E_X = (0.17 \pm 0.26)$ MeV. Though, it is experimentally not excluded that the mass of the $X(3872)$ is greater than the $\bar{D}^0 D^{*0}$ threshold, it is most likely that the molecule is bound.

In XEFT, contact interactions account for the short-distance degrees of freedom and long-range interactions are mediated by pion exchanges, similar to the Kaplan-Savage-Wise (KSW) theory for nucleon-nucleon (NN) interactions.[15] The pion exchanges are included perturbatively. This procedure fails in KSW theory due unnaturally large next-to-next-to-leading order (NNLO) coefficients, coming from the iteration of spin-tensor forces, and a rather large expansion factor for the perturbative inclusion of pions.[16] For XEFT, the expansion factor is much smaller and it is to be expected that it can compensate for unnaturally large coefficients at NNLO, if these occur.

The one-pion exchange sets a natural energy scale for the system, $m_\pi^2/(2M_{DD^*}) \approx 10$ MeV, where M_{DD^*} is the reduced mass of the D^0 and D^{*0} mesons. The binding energy is small compared to this energy scale, which leads to universal properties for the $X(3872)$, i.e. that low energy observables are mainly dependent on the small binding energy. Corrections to this universality can be calculated within XEFT.

We obtained analytical expressions for the binding energy and the scattering length in dependence on the light quark masses to NLO.[17] Our results are depicted in Figs. 1(a) and 1(b), respectively.

We plotted against the squared pion mass, which is proportional to the light quark masses according to the Gell-Mann-Oakes-Renner relation.[18]

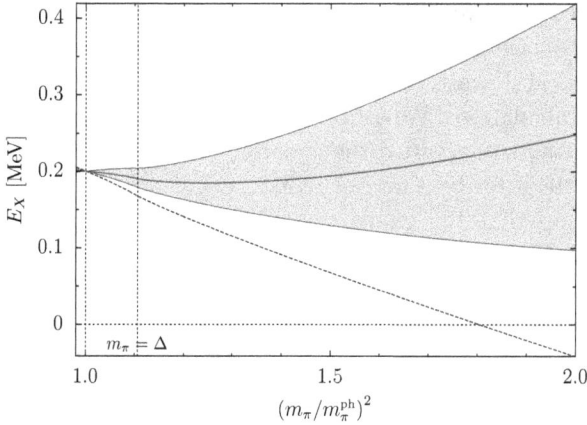

(a) Binding energy in dependence on the light quark masses

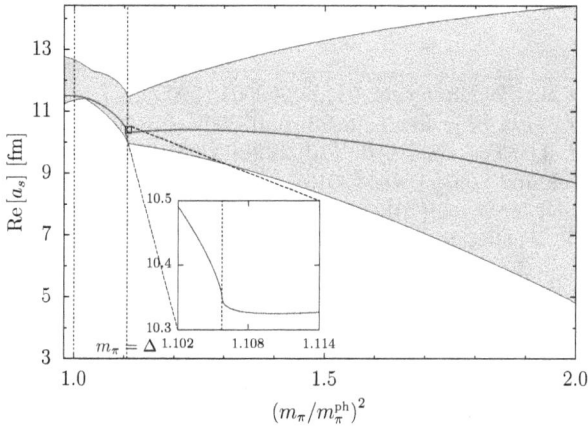

(b) Scattering length in dependence on the light quark masses

Fig. 1.   The solid, thick curves belong to the scenario where only the LO contact inter-action and pion exchanges and no NLO contact interactions are considered. The bounds are acquired by varying the NLO coupling constants in their natural ranges and max-imizing the width of the error band. The binding energy is fixed at the physical pion mass to $E_X = 0.2$ MeV.

Our results show that it is most likely that the X is bound for high quark masses and should be observable on the lattice.

We further found that a quark mass dependent contact interaction at NLO is essential to ensure that observables are renormalization scale independent, in analogy to KSW theory.[15] Another NLO contact interaction is required by power counting arguments. This introduces two unknown NLO coupling constants. While one of them can in principle be determined from experimental scattering data, the other one can only be acquired from lattice calculations. We estimated natural ranges for the values of the coupling constants and varied the coupling constants within these ranges, determining the error bars for our plots.

It shows up that the quark mass dependence of the binding energy and the scattering length is predominantly determined by the quark mass dependence of the pion exchanges and the NLO contact interactions. This requires chiral extrapolations for the axial coupling constant for the pions to the $D^{(*)}$ mesons and the pion decay constant.[19,20] We further considered chiral extrapolations for the masses of the $D^{(*)}$ mesons,[21] which play a minor role for the binding energy, but are important when considering the mass of the X.

References

1. S. Choi et al., *Phys.Rev.Lett.* **91**, p. 262001 (2003).
2. D. Acosta et al., *Phys.Rev.Lett.* **93**, p. 072001 (2004).
3. R. Aaij et al., *Phys. Rev. Lett.* 110, **222001** (2013).
4. S. Prelovsek and L. Leskovec (2013).
5. P. Wang and X. Wang (2013).
6. V. Baru, E. Epelbaum, A. Filin, C. Hanhart, U. G. Meißner et al. (2013).
7. S. Fleming, M. Kusunoki, T. Mehen and U. van Kolck, *Phys.Rev.* **D76**, p. 034006 (2007).
8. F. E. Close and P. R. Page, *Phys.Lett.* **B578**, 119 (2004).
9. S. Pakvasa and M. Suzuki, *Phys.Lett.* **B579**, 67 (2004).
10. M. Voloshin, *Phys.Lett.* **B579**, 316 (2004).
11. C.-Y. Wong, *Phys.Rev.* **C69**, p. 055202 (2004).
12. E. Braaten and M. Kusunoki, *Phys.Rev.* **D69**, p. 074005 (2004).
13. E. S. Swanson, *Phys.Lett.* **B588**, 189 (2004).
14. J. Beringer et al., *Phys.Rev.* **D86**, p. 010001 (2012).
15. D. B. Kaplan, M. J. Savage and M. B. Wise, **B424**, 390 (1998).
16. S. Fleming, T. Mehen and I. W. Stewart, *Nucl.Phys.* **A677**, 313 (2000).
17. M. Jansen, H. W. Hammer and Y. Jia (2013).
18. M. Gell-Mann, R. Oakes and B. Renner, *Phys.Rev.* **175**, 2195 (1968).
19. J. Gasser and H. Leutwyler, *Annals Phys.* **158**, p. 142 (1984).
20. D. Becirevic and F. Sanfilippo, *Phys.Lett.* **B721**, 94 (2013).
21. F.-K. Guo, C. Hanhart and U.-G. Meißner, *Eur.Phys.J.* **A40**, 171 (2009).

Hyperfine structure of ground-state nucleons in chiral quark model

Duojie Jia*, Wen-Bo Dang and Xing-Wen Zhao

Institute of Theoretical Physics, College of Physics and Electronic Engineering, Northwest Normal University, Lanzhou 730070, China
E-mail: jiadj@nwnu.edu.cn

Hyperfine structure of ground-state nucleons is studied by checking the isospin breaking effect due to the non-zero differences of mass and electromagnetic interaction between the up and down quarks. It is shown using chiral nonlinear quark model that the isospin breaking corrections to the baryon mass are of the order of one percent relatively to hadronic energies. The computed mass splittings due to the hyperfine strong and the electromagnetic corrections are in good agreement with the recent data of the baryon mass splittings.

Keywords: Hyperfine structure; Isospin breaking; Baryon mass; nonlinear chiral quark.

1. Introduction

In the Standard Model, the isospin symmetry is explicitly broken by the non-degenerate quark mass and the electromagnetic(EM) interaction between the up and down quark, and both of these corrections are expected to have a comparable size of the order of one percent relatively to hadronic energies.[1] This is indicated in the observed mass splitting in the hadron multiplets. In spite of their smallness relative to the strong mass splitting among different multiplets (e.g., the baryon octet and decuplet), these contributions play a crucial role in hadronic and nuclear physics. For instance, the mass difference and the EM interaction of the quarks is the key to account for the nucleon splitting,[2–4] which would constitute an ab-initio proof of nuclear matter stability and an interesting topic in hadron spectroscopy. Furthermore, the understanding of these isospin-breaking effects helps the fine determination of the light quark mass splitting which are fundamental for the standard model.

Recently, the QCD component of the mass splitting of the baryon octet due to up-down (and strange) quark mass differences is determined with lattice simulation in terms of the kaon mass splitting.[2,5] The further computation of the isospin corrections to decay constants and hadronic matrix elements requires more theoretical work to define properly these quantities in QCD and QED.[6] The role of isospin breaking is also emphasized in forming the bound state of X(3872).[7]

In this paper, we study the effect of isospin-symmetry breaking by computing the electromagnetic component of the mass splitting due to the non-degenerate mass and electric charge between the up-down quarks. We find using the chiral quark model with nonlinear pionic interaction that both of these corrections are of the order of 1% relatively to hadronic energies. The computed mass splitting due to the EM interaction and the non-degenerate mass, when adding the QCD component of hyperfine splitting, agrees well with the experimental values in baryon spectroscopy.

2. Quark Distribution in Baryon Multiplets

For baryon multiplets, the electromagnetic effects of and the mass differences between the light quarks account for the mass differences among the members of the multiplets. Before discussing such an EM contribution, we employ the chiral nonlinear quark model[1] to consider the quark distribution in baryons at low energy scale ($\sim 1 GeV$)

$$\mathcal{L}^{\chi QM} = \bar{q}[i\gamma^\mu \partial_\mu + \gamma^\mu v_\mu + \gamma^\mu \gamma^5 a_\mu - (m_q + S)U_5 - \gamma^0 V_G]q \\ + \mathcal{L}^{Skyrme} \tag{1}$$

with the pion dynamics given by the Skyrme Lagrangian,[8]

$$\mathcal{L}^{Skyrme} = \frac{f_\pi^2}{4} tr(\partial_\mu U \partial^\mu U^\dagger) + \frac{1}{32e_s^2} tr[\partial_\mu U U^\dagger, \partial_\nu U U^\dagger]^2 \\ + \frac{f_\pi^2}{4} tr[m_\pi^2 (U + U^\dagger - 2)] \, . \tag{2}$$

The EM effect will be treated as a perturbation (at the order of $1/\Lambda_{CSB}$) subsequently. Here, $U = \xi^2 = \exp(i\tau \cdot \pi/f_\pi)$ stands for the nonlinear representation of the Goldstone bosons (π mesons) under the constraint $U^\dagger U = 1$, $U_5 = \exp(i\gamma^5 \tau \cdot \pi/f_\pi)$, $\{v_\mu, a_\mu\} = \{(\xi^\dagger \partial_\mu \xi \pm \xi \partial_\mu \xi^\dagger)/2\}$, and f_π is the pion decay constant, and e_s the self-coupling of the nonlinear pion. We note that in the model (1) the confinement is put in phenomenologically, and chiral symmetry is implemented by construction.[1]

The static Hamiltonian associated with (1) is

$$H^{\chi QM} = \int d^3x q^\dagger [\alpha \cdot (\mathbf{p} + i a \gamma^5) + \gamma^0 (m_q + S U_5) + V_G]q \\ + H^{Skyrme}, \tag{3}$$

with $\mathbf{p} = -i\nabla$ the momentum operator, $\mathbf{a} \equiv i\nabla\pi/f_\pi$, and H^{Skyrme} is the Skyrme Hamilton. We apply, for simplicity, the hedgehog ansatz for the chiral field, $U(r) = \exp[iY(r)\tau \cdot \hat{r}]$, where r originated at the center of baryons. Taking the wavefunction of the (valence) quark to be $q = \frac{N}{r}(G(r), -iF(r)\sigma \cdot \hat{r})^t y_{ljm}(\theta\varphi)\chi_f$ with $N^2 \equiv 1/[\int dr(G^2 + F^2)]$, y_{ljm} the Pauli spinor and χ_f the flavor wavefunction, one finds for (3)

$$H^{\chi QM} = N^2 \int dr \left\{ F\frac{dG}{dr} - G\frac{dF}{dr} + \frac{2\kappa}{r}GF + (m_q + S)\cos Y(G^2 - F^2) \right\} \\ + H^A + H^{Skyrme},$$

with $-\kappa$ the eigenvalue of the operator $K = \gamma^0[\Sigma \cdot (\mathbf{r} \times \mathbf{p}) + 1]$ corresponding to the eigenstate y_{ljm}, and $H^A = -(g_A L^2 N^2/2)(\chi_f^\dagger \sigma_r \chi_f) \int dr d\Omega - y_{ljm}|^2 Y_x(G^2 - F^2)$. The eqnarray of motion for (3) with the normalization condition of the quark wavefunction, for S-state($l = 0$), is

$$\frac{dG}{dz} + \frac{\kappa}{z}G = [\varepsilon_q + L(m_q + S)\cos Y - LV_G]F$$
$$-\frac{dH}{dz} + \frac{\kappa}{z}F = [\varepsilon_q - L(m_q + S)\cos Y - LV_G]G$$
$$\left(1 + \frac{2\sin^2 Y}{z^2}\right)Y_{zz} + \frac{2\sin(2Y)}{z}\left(Y_z^2 - 1 - \frac{\sin^2(Y)}{z^2}\right) \qquad (4)$$
$$= e^2 \left[\left(\frac{Lm_\pi}{e_s}\right)^2 - \frac{LN^2}{4\pi}(Lm_q + LS)\frac{(G^2 - F^2)}{z^2} \right]\sin Y$$

where $\varepsilon_q = E_q L$ is the energy eigenvalue of q, and $z \equiv r/L \equiv ef_\pi r$. Here, the QCD confining interaction is chosen to be of the Cornell form

$$LS \equiv L(r/a^2) = L_a^2 z,$$
$$LV_G \equiv L(\alpha/r) = \alpha/z, \qquad (5)$$

with $L_a = 1/(e_s f_\pi a)$, a is the confining scale and α the flavor depended strong coupling.

The asymptotic solution to (4) $G \sim e^{-L_a^2 z^2/2} \sim -F$ and $F \sim e^{-2\sqrt{B}L^2 z}/z$ at $z \to \infty$ and that $G = \sqrt{z}Y_{\kappa+1/2}(\varepsilon z) \sim \sqrt{\frac{2\varepsilon}{\pi}}z, H = -zJ_{\kappa+3/2}(\varepsilon z) \sim -\sqrt{\frac{2\varepsilon}{\pi}}z^{3/2}, F = \pi - Az$ at $z \to 0$ will be applied to find the numerical solution of the quark wavefunction. For the chiral angle profile, we utilize, as an approximation,[9,10] the parameterized Skyrmion profile

$$Y(r) \simeq 4w \arctan[\exp(-c_0 z)] + (1 - w)\pi \left[1 - \left(\frac{\sinh^2(dz)}{a_0^2 + \sinh^2(dz)}\right)^{-1/2}\right], \qquad (6)$$

in which the parameters $\{c, a_0, w, d\}$ are to be fixed by optimizing H^{Skyrme} (with Neilder-Mead algorithm) as it is done in Ref. 10. We plot the up

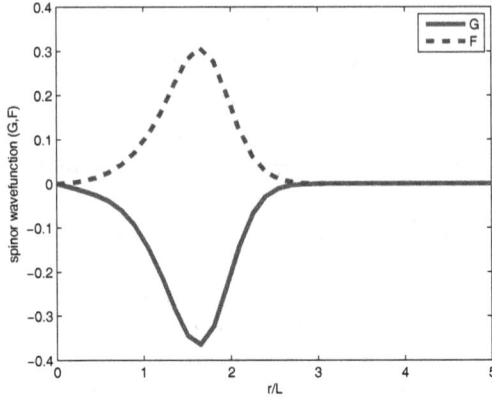

Fig. 1.   The quark wavefunction.

quark wavefunction in Fig. 1 for the parameters

$$f_\pi = 93 MeV, e_s = 2.80, m_\pi = 137.6 MeV;$$
$$\alpha = 0.8, a = 1/0.6676 GeV^{-1}, m_u = 361.75, m_d = 364.25. \tag{7}$$

The constants for the chiral angle profile optimized are

$$\{a_0, c, d, w\} = \{0.620, 0.904, 1.283, 1.174\}. \tag{8}$$

To avoid the non-conservation of particle number due to the sea-quark effect in the relativistic model it is convenient for a quark state to use the nonrelativistic wavefunction given in terms of the original spinor wavefunction,

$$\psi_q(\mathbf{x}) = \frac{NG(r)}{r}[1 + F(r)^2/G(r)^2]^{1/2}, \tag{9}$$

which satisfies the normalization $\int |\psi_q(r)y_{ljm}|^2 d^3x = 1$, and plotted in Fig. 2.

We note that in solving (4) for the quark wavefunction we take the isospin-symmetry approximation so that $H^A \sim \chi_f^\dagger \sigma_r \chi_f \sim 0$ for the two flavor case, as it is suppressed by $1/\Lambda_{CSB}$. The correction to the quark wavefunction due to the isospin breaking can be given separately by taking H^A (order of $1/\Lambda_{CSB}$) into account perturbatively, leading to a tiny correction to baryon mass suppressed by $1/\Lambda_{CSB}^2$.

The average sizes of baryons, when defined as $R_0 = Y^{-1}(\pi/2)$, has a typical value $R_0 \simeq 0.8 fm$, whereas the size of the valance quark distribution

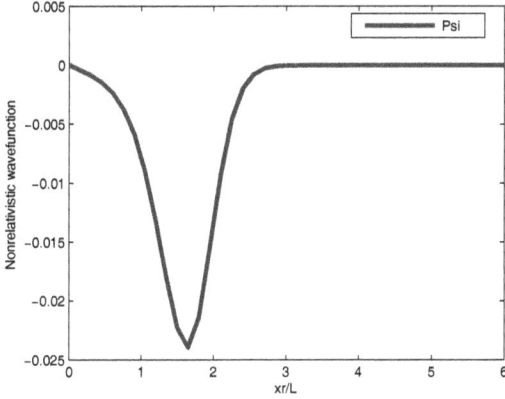

Fig. 2. The nonrelativistic wavefunction of quark.

$|\psi_{q_i}(\mathbf{x})|^2$ in baryon depends on the details of the spatial quark wavefunction $\psi_{q_i}(\mathbf{x})$, in which the effects due to inter-quark correlation have been included in the sense of the large-N_c mean-field approximation (N_c is the number of colors).[11,12] We will use in the following sections the solutions to (4) for the up-down quarks to compute the EM energy correction to the baryon mass.

3. Coulomb Energy Differences in Baryons

According to the Standard Model, the mass differences in hadron multiplets have the origin due to the QED effects of quark-quark interaction as well as the mass differences between the light quarks. We take such a mass difference as input so that the quark distribution becomes different slightly between u and d flavors. To explore the EM contribution to the isospin breaking, we first consider the coulomb energy associated with the electric energy between pairs of quarks in baryons. This energy will be roughly of order of $e^2/R_0 \simeq 2MeV$, with $R_0 \simeq 0.8fm$ being the size of a low-lying baryon.

The above estimate for the electric (Coulomb) energy between pairs (ij) of quark i and j in baryon can be improved by evaluating the following expectation value in term of the confined quark wavefunction

$$\Delta E_{ij}^{(c)} = \langle q_i q_j | \frac{e_i e_j}{4\pi\epsilon_0 |\mathbf{x}_1 - \mathbf{x}_2|} | q_i q_j \rangle \tag{10}$$

where $e_i (i = u, d)$ stands for the quark charges in the unit of electric charge $|e|$, and $q_i q_j$ (the flavor part suppressed temporarily) the two-body quark wavefunction of the coordinate $\mathbf{x}_1$ and $\mathbf{x}_2$. We have taken $q_i q_j$ to be roughly equal to $\psi_{q_i}(\mathbf{x}_1)\psi_{q_j}(\mathbf{x}_2)y_{ljm}(\Omega_1)y_{ljm}(\Omega_2)$ since the dominate effect of two-body interaction is already included in the chiral mean-field interaction and the confining interaction in (1), as it is the case when the Goldstone field introduced as a classical large-N_c mean-field field (N_c is the number of colors).[11,12] The situation here is then very similar to the effects of electron-electron Coulomb interaction in the helium atom, and (in the natural unit where $4\pi\epsilon_0 = 1$) one can rewrite (10) as (for S-state)

$$
\begin{aligned}
\Delta E_{ij}^{(c)} &= e_i e_j \int\int d\mathbf{x}_1 d\mathbf{x}_2 \, y_{ljm}(\Omega_1) y_{ljm}(\Omega_2) \frac{|\psi_{q_i}(\mathbf{x}_1)|^2 |\psi_{q_j}(\mathbf{x}_2)|^2}{|\mathbf{x}_1 - \mathbf{x}_2|} \\
&= \frac{e_i e_j}{2} \int |\psi_{q_i}(\mathbf{x}_1)|^2 r_1 dr_1 \int |\psi_{q_j}(\mathbf{x}_2)|^2 r_2 dr_2 (r_1 + r_2 - |r_1 - r_2|) \\
&= e_i e_j \int_0^\infty r_1 dr_1 |\psi_{q_i}(r_1)|^2 \left[\int_0^{r_1} |\psi_{q_j}(r_2)|^2 r_2^2 dr_2 \right. \\
&\quad \left. + r_1 \int_{r_1}^\infty |\psi_{q_j}(r_2)|^2 r_2 dr_2 \right].
\end{aligned}
\tag{11}
$$

For the integration in (11), we need the relations

$$
\int |\psi_{q_i}(r)|^2 r^n dr = \frac{\int dr r^{n-1}[G^2 + F^2]}{\int dr [G^2 + F^2]}.
$$

The total electric (Coulomb) energy in baryon will then be

$$
\Delta E^{(c)} = \sum_{i,j} \Delta E_{ij}^{(c)} = \alpha \sum_{i,j} e_i e_j C_{ij},
\tag{12}
$$

where (when (9) used)

$$
\begin{aligned}
C_{ij} \equiv (N_i^2 N_j^2 L) \int_0^\infty \frac{dz_1}{z_1} [G_i(z_1)^2 + F_i(z_1)^2] \\
\times \left[\int_0^{z_1} [G_j(z_2)^2 + F_j(z_2)^2] dz_2 + z_1 \int_{z_1}^\infty [G_j(z_2)^2 + F_j(z_2)^2] \frac{dz_2}{z_2} \right],
\end{aligned}
\tag{13}
$$

with $(N_i G_i, N_i F_i)$ being the radial part of the wavefunction of the quark i in (9).

We apply now the formula (12) to the nucleons (proton and neutron) and the delta hyperons. Employing the SU(6) spin-flavor wavefunction of quark model, the electric mass differences for these baryons are

Table 1. The inter-quark factors and EM energy(MeV).

pairs(ij) baryons	C_{ij}	H_{ij}	$\Delta E_{ij}^{(c)}$	$\Delta E_{ij}^{(M)}(\uparrow\uparrow)$	$\Delta E_{ij}^{(M)}(\uparrow\downarrow)$
uu	2.6851	0.0042	8.7	0.26	-0.77
ud	2.6853	0.0010	4.3	0.06	-0.18
dd	2.6852	-0.0021	-2.2	-0.13	0.38

$(e_u = 2/3, e_d = -1/3)$

$$\Delta E^{(c)}(\text{p}) = \alpha \left(\tfrac{4}{9}\right) [C_{uu} - C_{ud}] = \Delta E^{(c)}(\Delta^+)$$
$$\Delta E^{(c)}(\text{n}) = \alpha \left(\tfrac{1}{9}\right) [C_{dd} - 4C_{ud}] = \Delta E^{(c)}(\Delta^0)$$
$$\Delta E^{(c)}(\Delta^{++}) = \alpha \left(\tfrac{4}{3}\right) C_{uu} \tag{14}$$
$$\Delta E^{(c)}(\Delta^-) = \alpha \left(\tfrac{1}{3}\right) C_{dd}.$$

We calculate C_{ij} for the up-down quarks and list them in the Table 1, using the numerical solutions to (4) that obtained in the Sec. 2. The electric mass corrections are listed in the Table 2.

4. Magnetic Energy Differences

Another dominate isospin-breaking contribution to mass difference in nucleons arises from the hyperfine interaction(magnetic-moment interaction) between the quark pairs in the baryons. When the nucleons are in S-state, such a contribution can be written as the spin-spin interaction energy

$$\Delta E_{ij}^{(M)} = \langle q_i(\mathbf{x}_1)| \frac{8\pi e_i e_j}{3m_i m_j} \mathbf{S}_i \cdot \mathbf{S}_j \delta(\mathbf{x}_1 - \mathbf{x}_2)|q_j(\mathbf{x}_2)\rangle \tag{15}$$

$$= \frac{8\pi\alpha e_i e_j}{3m_i m_j} h_{ij}\sigma_i \cdot \sigma_j$$

where $\alpha = e^2/(\hbar c)$ is the QED hyperfine constant, $h_{ij} \equiv |\psi_{q_i q_j}(0)|^2$ the amplitude of two-body wavefunction $\psi_{q_i q_j}(\mathbf{r})$ at the origin of the relative coordinate $(\mathbf{r} = \mathbf{x}_1 - \mathbf{x}_2 = 0)$, namely, the probability for finding the quark q_i and q_j at the coinciding coordinate,

$$h_{ij} \equiv |\langle q_i(\mathbf{x})|q_j(\mathbf{x})\rangle|^2 = \int d\mathbf{x}|\psi_{q_i}(\mathbf{x})|^2|\psi_{q_j}(\mathbf{x})|^2. \tag{16}$$

Then, the total magnetic energy among the pairs of quark (i, j) in baryon is

$$\Delta E^{(M)} = \sum_{i,j} \Delta E_{ij}^{(M)} = \frac{8\pi\alpha}{3} \sum_{i,j} H_{ij}\sigma_i \cdot \sigma_j, \tag{17}$$

where

$$H_{ij} = \frac{e_i e_j}{m_i m_j} h_{ij}. \tag{18}$$

Applying (17) to the nucleons and the delta hyperons, we calculate the magnetic coupling H_{ij} and the magnetic energy$(\Delta E_{ij}^{(M)})$ and the corresponding magnetic-energy correction to the baryon masses, as listed in Table 1.

Similar to the computation of the Coulomb energy in Section 3, we use (17) to compute the magnetic energy of the nucleons and the delta hyperons. In the case of the nucleons, the spin-spin product in (17) for a specific pair (ij) should be the average value of all possible spin configurations $(s_i s_j)$:

$$\langle \sigma_i \cdot \sigma_j \rangle = \frac{1}{3} \sum_{s_i s_j} w(s_i s_j) 2[S(S+1) - s_i(s_i+1) - s_j(s_j+1)],$$

in which the probability weights $w(s_i s_j)$ are extracted from the SU(6) spin-flavor wavefunction of the proton and neutron. For the spin configuration $(s_i s_j)$, they can be listed as

$$\text{octets}: proton \left\{ \begin{array}{ccc} w(s_i s_j) & uu & ud \\ \uparrow\uparrow & 2/3 & 1/3 \\ \uparrow\downarrow & 1/3 & 5/3 \end{array} \right\} ; neutron \left\{ \begin{array}{ccc} w(s_i s_j) & ud & dd \\ \uparrow\uparrow & 1/3 & 2/3 \\ \uparrow\downarrow & 5/3 & 1/3 \end{array} \right\}.$$

In the case of decuplets, however, one always has $\sigma_i \cdot \sigma_j = 1$ since the spin of all pairs are parallel.

Put together the electric contribution (12) and magnetic contribution (17) to the baryon mass, one has

$$\Delta E^{EM} = \Delta E^{(c)} + \Delta E^{(M)}$$
$$= \alpha \sum_{i,j} \left[e_i e_j C_{ij} + \frac{8\pi H_{ij}}{3} \sigma_i \cdot \sigma_j \right]. \tag{19}$$

Another component of mass splitting has the origin of the strong interaction(QCD), which can be given by the similar spin-spin interaction to (17), but with the prefactor A of order of the squared mass of the constituent quark(m_q^2),

$$A \sum_{i,j} \frac{\mathbf{S}_i \cdot \mathbf{S}_j}{m_i m_j}, \tag{20}$$

in which A will be chosen to be $(2m_u)^2 \cdot 50 MeV$.[13] Up to an additional mass $M_0 = 939 MeV$ which is fitted phenomenologically, we find a formula for the mass differences for baryons

$$m(\text{baryon}) - M_0 = A \frac{\mathbf{S}_i \cdot \mathbf{S}_j}{m_i m_j} + \alpha \sum_{i,j} \left[e_i e_j C_{ij} + \frac{8\pi H_{ij}}{3} \sigma_i \cdot \sigma_j \right]. \tag{21}$$

Table 2. The EM energy differences.

Baryons	$p - n$	p	n	$\Delta^{++} - \Delta^0$	$\Delta^+ - \Delta^0$	$\Delta^- - \Delta^0$
ΔE^{EM}	2.08	1.4	-0.68	3.03	5.3	1.4
$\Delta E^{EM}(^5)$	0.383(68)					

Evidently, one can see that the EM component of the mass splitting in (21) is suppressed by the QED fine structural constant $\alpha = 1/137$, which is of the order of one percent relatively to hadronic(baryon) mass. The numerical results of the formula (19) and (21) are listed in the Table 2.

5. Summary

We studied the hyperfine structure of ground-state nucleons by computing the isospin-breaking effects due to the non-degenerate mass of and electromagnetic interaction between u and d quarks. We show within the framework of the chiral nonlinear quark model that these isospin-breaking corrections are of the order of one percent relatively to hadronic(baryonic) energies. The mass splitting due to the hyperfine strong and electromagnetic corrections are found to be in good agreement with the recent data.

Acknowledgments

D. J thanks A. Hosaka, Jun He, Xiang Liu, and Qiang Zhao for useful discussions. This work is supported by the National Natural Science Foundation of China (No.11265014) and (No.10965005).

References

1. H. Georgi, *Weak interactions and modern particle theory*, Dover. Inc. Mineola (N.Y.)2009.
2. R. Horsleya, J. Najjarb et al., (QCDSF-UKQCD Collaboration), Phys. Rev. D86(2012)114511.
3. P. Hagler, Phys. Rept. 490(2010) 49-175.
4. J. Gasser, H. Leutwyler, Ann. Phys. (NY)158(1984)142; Phys. Rep. 87 (1982)77
5. Tom Blum and Ran Zhou et al., Phys. Rev. D82 (2010)094508.
6. J. Gasser and G. R. S. Zarnauskas. On the pion decay constant. Phys. Lett. B 693(2) pp. 122–128 (2010).
7. NingLi and Shi-Lin Zhu, Phys. Rev. D.86.074022_2012.
8. T. H. R. Skyrme, Proc. R. Soc. A 260(1961)127; Nucl. Phys. 31(1962)556.
9. M. F. Atiyah and N. S. Manton, Commun. Math. Phys. 152(1993)391.
10. Duojie Jia, Xiao-Wei Wang, Mod. Phys. Lett. A26(2011)557-565.
11. E. Witten, Nucl. Phys. B160 (1979) 57 .
12. D.Diakonov, V.Petrov and A.A. Vladimirov, Phys. Rev. D88 (2013)074030.
13. S. Gasiorowicz and J.L. Rosner, Am. J. Phys. 49(1981)954.

The high order chiral Lagrangian

Shao-Zhou Jiang

College of Physical Science and Technology, Guangxi University,
Nanning, Guangxi, P. R. China
E-mail: jsz@gxu.edu.cn

Qing Wang

Department of Physics, Tsinghua University, Beijing, P. R. China
E-mail: wangq@mail.tsinghua.edu.cn

We obtain the full U group chiral Lagrangian to the order p^6 order, and check the existing results. We find one more linear relation in p^4 order at U group, and some more linear relations in p^6 order at SU group with tensor sources.

Keywords: Chiral lagrangian; QCD; spontaneous symmetry breaking.

1. Introduction

Chiral perturbation theory is effective widely in the low-energy QCD, here we only mention about the construction of the pseudoscalar mesons chiral Lagrangian. Until now, under the SU group, the p^2 order,[1] p^4 order[2,3] and p^6 order chiral Lagrangian have been obtained, including the normal[4-6] (SU_n) and anomalous[4,7,8] parts (SU_a), and tensor sources[9,10] (SU_t). Under the U group, the p^4 order chiral lagrangian was also obtained.[11] Table 1 is a summary. We list the number of independent terms in each given case. But the p^6 order U group results do not exist. Our aim is to obtain the full U group chiral Lagrangian in the p^6 order.

Table 1. Numbers of independent operators.

order	p^2			p^4			p^6		
group	n	3	2	n	3	2	n	3	2
SU_n	2	2	2	$11+2$	$10+2$	$7+3$	$112+3$	$90+4$	$52+4$
SU_a	0	0	0	Wezz	Zumino	term	24	23	$5+8$
U	6	6	6		$50+7$				
SU_t	0	0	0	4	4	4	$96+2$	$90+2$	$63+2$

2. Symbols and Definitions

Now, we come to QCD Lagrangian Eq.(1),

$$\mathcal{L} = \mathcal{L}^0_{QCD} + \bar{q}(\slashed{v} + \slashed{a}\gamma_5 - s + ip\gamma_5 + \sigma_{\mu\nu}\bar{t}^{\mu\nu})q - \frac{\theta}{16\pi^2}\text{tr}_c(G_{\mu\nu}\tilde{G}^{\mu\nu}). \quad (1)$$

$\mathcal{L}^0_{QCD}$ is the original QCD Lagrangian, including vector(v), axial-vector(a), scalar(s), pseudo-scalar(p) and tensor source($\bar{t}$), and vacuum term(θ). The lightest nonet meson fields is usually collected as $U(x) = e^{i\phi(x)/F_0}$, where $\phi = \boldsymbol{\lambda} \cdot \boldsymbol{\pi}, \lambda_0 = \sqrt{2/n_f}I$ and $\lambda_i(i = 1,\ldots,8)$ are the Gell-Mann matrices of $SU(n_f)$. If we set quark mass to zeros, there exists a $U_L(n_f) \times U_R(n_f)$ chiral symmetry. Under these chiral rotation, U fields transforms as $U \to g_R U g_L^\dagger$. To construct the chiral lagrangian, one often introduces a field $u, u^2 = U$, with a compensator field h. Under the chiral rotation, u transform as $u \to u = g_R u h^\dagger = h u g_L^\dagger$. The mesonic chiral Lagrangian can be constructed by some building blocks O, which is transform as $O \to hOh^{-1}$ or invariant under the chiral rotations. The building blocks are in Eq.(2),

$$u^\mu = i\{u^\dagger(\partial^\mu - ir^\mu)u - u(\partial^\mu - il^\mu)u^\dagger\}, \quad \chi_\pm = u^\dagger\chi u^\dagger \pm u\chi^\dagger u,$$

$$h^{\mu\nu} = \nabla^\mu u^\nu + \nabla^\nu u^\mu, \quad f_\pm^{\mu\nu} = uF_L^{\mu\nu}u^\dagger \pm u^\dagger F_R^{\mu\nu}u, \quad t_\pm^{\mu\nu} = u^\dagger t^{\mu\nu}u^\dagger \pm ut^{\mu\nu\dagger}u,$$

$$\nabla^\mu\hat{\theta} = \partial^\mu\hat{\theta} + 2i\langle a^\mu\rangle, \quad X = \langle\ln U\rangle + \hat{\theta}. \quad (2)$$

with

$$r^\mu = v^\mu + a^\mu, l^\mu = v^\mu - a^\mu, \chi = 2B_0(s + ip),$$

$$t^{\mu\nu} = \frac{1}{2}\bar{t}^{\mu\nu} - \frac{i}{4}\epsilon^{\mu\nu\lambda\rho}\bar{t}_{\lambda\rho}, \hat{\theta} = i\theta,$$

$$F_R^{\mu\nu} = \partial^\mu r^\nu - \partial^\nu r^\mu - i[r^\mu, r^\nu], F_L^{\mu\nu} = \partial^\mu l^\nu - \partial^\nu l^\mu - i[l^\mu, l^\nu]. \quad (3)$$

Because mesonic chiral Lagrangian is C, P and hermitian invariant, we also collect all the building blocks C. P and hermitian conjugate properties in Table 2.

Table 2. Building blocks and their properties.

O	P	C	h.c.
u^μ	$-u_\mu$	$(u^\mu)^T$	u^μ
$h^{\mu\nu}$	$-h_{\mu\nu}$	$(h^{\mu\nu})^T$	$h^{\mu\nu}$
$\chi_\pm$	$\pm\chi_\pm$	$(\chi_\pm)^T$	$\pm\chi_\pm$
$f_\pm^{\mu\nu}$	$\pm f_{\pm\mu\nu}$	$\mp(f_\pm^{\mu\nu})^T$	$f_\pm^{\mu\nu}$
$t_\pm^{\mu\nu}$	$\pm t_{\pm\mu\nu}$	$-(t_\pm^{\mu\nu})^T$	$\pm t_\pm^{\mu\nu}$
$\nabla_\mu\hat{\theta}$	$-\nabla_\mu\hat{\theta}$	$\nabla_\mu\hat{\theta}$	$-\nabla_\mu\hat{\theta}$
X	$-X$	X	X

3. Construction of chiral Lagrangian

To construct chiral Lagrangian, we use the conventional power counting rules, $U, \hat{\theta}, X \sim O(p^0), v_\mu, a_\mu \sim O(p^1), \chi, t_{\mu\nu} \sim O(p^2)$, and the covariant derivative(∇) count as $O(p^1)$. It is defined as

$$\nabla^\mu O = \partial^\mu O + [\Gamma^\mu, O], \quad \Gamma^\mu = \frac{1}{2}\{u^\dagger(\partial^\mu - ir^\mu)u + u(\partial^\mu - il^\mu)u^\dagger\}. \quad (4)$$

The Lagrangian can be write as

$$\mathcal{L} = \sum_i C_i(X)O_i \quad (5)$$

$C_i(X)$ is the function of X, O_i are some operators in $O(p^{2n})$ order. Because X is p^0 order and its parity is odd, when we write down the Lagrangian, O_i can be odd parity.

The p^0 and p^2 order Lagrangian were found very early,[11] the number of terms is seven. When we construct higher order Lagrangian, it need to consider the following relations to remove all linear dependent terms:

(1) Ignoring a total integration, we can remove one covariant derivative to another operator, such as $\langle \nabla^\mu \nabla^\nu AB \rangle = -\langle \nabla^\nu A \nabla^\mu B \rangle$.
(2) With the equations of motion, Eq.(22) in Ref.,[11] we can remove $\nabla^\mu u_\mu$.
(3) With Bianchi identity, Eq.(6), we can change the index of ∇ and Γ.

$$\nabla^\mu \Gamma^{\nu\lambda} + \nabla^\nu \Gamma^{\lambda\mu} + \nabla^\lambda \Gamma^{\mu\nu} = 0, \quad (6)$$

(4) With tensor relations,[9] $\epsilon_{\mu\nu\lambda\rho}t_\pm^{\lambda\rho} = 2it_{\mp\mu\nu}$, more than one t_- terms can be removed, and $t_\pm$ can be exchanged.
(5) If the Lagrangian include the ϵ parameter, Schouten identity may be used

$$\epsilon^{\mu\nu\lambda\rho}A^\sigma - \epsilon^{\sigma\nu\lambda\rho}A^\mu - \epsilon^{\mu\sigma\lambda\rho}A^\nu - \epsilon^{\mu\nu\sigma\rho}A^\lambda - \epsilon^{\mu\nu\lambda\sigma}A^\rho = 0. \quad (7)$$

(6) In a given flavor n_f, Cayley-Hamilton relation, Eq.(3.1), (3.3) in Ref. 5, gives an additional relation.
(7) Finally, to pick up the non-meson fields, the contact terms need to be constructed. To construct contact terms, we come to Left-Right basis, $F_{L,R}^{\mu\nu}, \chi^{(\dagger)}, t^{(\dagger)\mu\nu}, X, \nabla^\mu \hat{\theta}$ in Eq.(3). Because the building blocks blend u fields badly.

Construct the chiral Lagrangian is sounded easily. First, we pile the building blocks, to find all terms are invariant, under C, P and Hermitian conjugate transforms. Then pick out contact terms, and remove all linear depend terms. Non-contact terms are removed first. But the calculation is too complex, it need the help of a computer.

4. Results

We have repeat the all results to p^6 order in the existing references, to check our method and the existing results. p^0 and p^2 order results are confirmed. The p^4 order results are all most the same, but one more linear relation than Ref.,[11] $O_{29} = O_{28}$. It can be checked with Eq.(3) directly, or much simpler with Eq.(2). The full results are too long, and will be presented elsewhere.[12]

In p^6 order, about the SU group, we find only the tensor sources parts can be reduced. In n_f flavor, Y_{102} can be reduced, and other nine operators, $Y_{48}, Y_{49}, Y_{50}, Y_{71}, Y_{72}, Y_{73}, Y_{92}, Y_{106}, Y_{108}$

About the U group, we list the number of independent terms in Table 3. The full results will be presented elsewhere.[12]

Table 3. Numbers of U group results

Number	$P = +/-$	U(N)			SU(N)$_I$			SU(N)$_{II}$			SU(N)$_{III}$		
classification	P	n	3	2	n	3	2	n	3	2	n	3	2
$\theta = \bar{t}^{\mu\nu} = 0$	287/206	493	451	281	155	133	74	263	235	142	138	116	60
$\theta \neq 0, \bar{t}^{\mu\nu} = 0$	261/206	467	455	371	124	118	83	251	243	194	116	110	75
$\theta = 0, \bar{t}^{\mu\nu} \neq 0$	159/149	308	297	209	110	104	67	137	131	88	97	91	56
$\theta \neq 0, \bar{t}^{\mu\nu} \neq 0$	64/59	123	123	108	51	51	43	52	52	44	42	42	34
total	790/675	1465	1399	1034	454	420	281	705	663	470	395	361	227

References

1. S. Weinberg, *Physica* **96A**, p. 327 (1979).
2. J. Gasser and H. Leutwyler, *Ann. Phys. (N.Y.)* **158**, p. 142 (1984).
3. J. Gasser and H. Leutwyler, *Nucl. Phys.* **B250**, p. 465 (1985).
4. H. W. Fearing and S. Scherer, *Phys. Rev. D* **53**, p. 315 (1996).
5. J. Bijnens, G. Colangelo and G. Ecker, *J. High Energy Phys.* **02**, p. 020 (1999).
6. C. Haefeli, M. A. Ivanov, M. Schmid and G. Ecker. arXiv:0705.0576.
7. T. Ebertshäuser, H. Fearing and S. Scherer, *Phys. Rev. D* **65**, p. 054033 (2002).
8. J. Bijnens, L. Girlanda and P. Talavera, *Eur. Phys. J. C* **23**, p. 539 (2002).
9. O. Catà and V. Mateu, *J. High Energy Phys.* **09**, p. 078 (2007).
10. S.-Z. Jiang, Y. Zhang and Q. Wang, *Phys. Rev. D* **87**, p. 094014 (2013).
11. P. Herrera-Siklódy, J. Latorre, P. Pascual and J. Taron, *Nucl.Phys.* **B497**, p. 345 (1997).
12. in preparation.

The $f_0(1790)$ and $f_0(1800)$ puzzle

K. P. Khemchandani, A. Martínez Torres, M. Nielsen, F. S. Navarra

Instituto de Física, Universidade de São Paulo, C.P 66318,
05314-970 São Paulo, SP, Brazil

D. Jido

Department of Physics, Tokyo Metropolitan University,
Hachioji, Tokyo 192-0397, Japan

A. Hosaka

Research Center for Nuclear Physics (RCNP), Mihogaoka 10-1,
Ibaraki 567-0047, Japan

E. Oset

Departamento de Física Teórica and IFIC, Centro Mixto Universidad de
Valencia-CSIC, Institutos de Investigación de Paterna,
Aptd. 22085, 46071 Valencia, Spain

This manuscript sheds light on the puzzle created by the finding of two resonances with same quantum numbers and very similar mass but which possess very different decay properties. One of them, $f_0(1790)$, has been found in the $J/\Psi \to \phi\pi^+\pi^-$ process while the other, $f_0(1800)$, in $J/\Psi \to \gamma\omega\phi$. As we shall discuss in this manuscript, our studies show that these two states are distinct to each other but only one of them is a new state, while the other one is the manifestation of the known $f_0(1710)$. We also provide an explanation for the different decay properties of these states.

Keywords: Scalar resonances; few-body systems; decay channels.

1. Introduction

Existence of unstable hadrons or resonances in the intermediate energy region makes the study of hadron spectroscopy challenging from experimental point of view. Usually, a resonance is identified in experimental data by finding an enhancement in the cross sections with significant statistics. However, it can be very tricky to judge if the peak position in the data corresponds to the mass of the resonance. Such difficulties can sometimes create confusions. We shall discuss one such case here. One of the puzzles appeared recently with the finding of the $f_0(1800)$ resonance in the $\phi\omega$ invariant mass spectrum in a recent experimental study of the $J/\Psi \to \gamma\phi\omega$ process by the BES collaboration.[1] The mass and the width of the resonance were given to be $M = 1795 \pm 7^{+23}_{-5}$ MeV and $\Gamma = 95 \pm 10^{+78}_{-34}$ MeV, respectively, in Ref. 1. Coincidently, a f_0 resonance with a very similar mass ($M = 1790^{+40}_{-30}$ MeV) was also found in a previous experiment carried out by the BES collaboration, on a different decay channel, $\phi\pi^+\pi^-$ of J/Ψ.[2] This f_0 was found in the $\pi\pi$ mass spectrum and was named as $f_0(1790)$ in Ref. 2. It was also found in Ref. 2 that the decay of $f_0(1790)$ to the $K\bar{K}$ channel is suppressed. This property clearly distinguishes $f_0(1790)$ found in Ref. 2 with the $f_0(1800)$ found in Ref. 1 which can easily decay to $K\bar{K}$ (through a kaon/anti-kaon exchange between ϕ and ω (see Refs. 3,4 for more details)) and to $f_0(1710)$ which is known to have large branching ratio for $K\bar{K}$. Thus, from these data it seems that two f_0 resonances with similar masses but opposite preferences to the decay channel $K\bar{K}$ exist. Our studies[4,5] show that the peak seen in the $\phi\omega$ spectrum is not a new resonance, that it is the manifestation of $f_0(1710)$, and that $f_0(1790)$ is a new resonance which can be understood as one arising from the $\pi\pi f_0(980)$ dynamics. We review some details of our studies[4,5] in the next section.

2. The Nature of f_0 Resonances in the 1700-1800 MeV Interval

There is a well known scalar, isoscalar resonance in the 1700-1800 MeV energy range, which is $f_0(1710)$. But the new findings of Refs. 1,2 show that there might be two more f_0's lying in this range. Let us first see that one of this new resonances is redundant since the signal in the $\phi\omega$ mass spectrum can be explained in terms of the tail of the $f_0(1710)$ resonance. In fact it was found in Ref. 6 that $f_0(1710)$ is dynamically generated in two vector systems, a finding which has been very useful in reproducing the $\phi\omega$ spectrum in Ref. 5 and which was missing in Ref. 7 where an attempt to explain the same data with $f_0(1710)$ was made and a peak structure with much weaker strength than data was found. We studied the $J/\Psi \to \gamma\phi\omega$ process in Ref. 5 using the formalism shown diagrammatically in Fig. 1. In this formalism a photon is radiated from the initial $c\bar{c}$ state (as in Ref. 8) which, then, as a SU(3) singlet decays into pairs of vectors which interact among themselves.

Within this formalism we obtain the $\phi\omega$ mass distribution shown in Fig. 2.

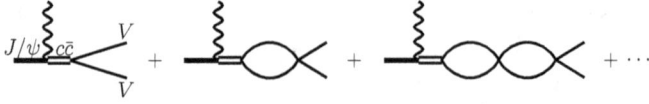

Fig. 1. Schematic representation of J/Ψ decay into a photon and one dynamically generated resonance.

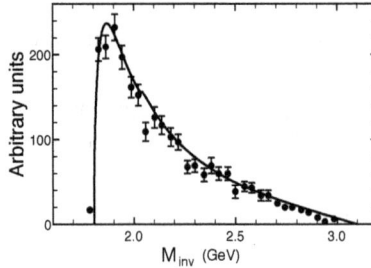

Fig. 2. The $\phi\omega$ invariant mass distribution calculated using Eq. (12) of Ref. 5. The data points, shown by filled circles, have been taken from Ref. 1.

Now, the resonance found in Ref. 2 can, certainly, not be related to the one found in the $\phi\omega$ invariant mass spectrum in Ref. 1 since the latter one must unavoidably decay to $K\overline{K}$ through a kaon exchange and the decay of the former one to $K\overline{K}$ is suppressed. This argument actually leads to the finding of a flaw in the interpretation of the peak seen in the $\phi\omega$ spectrum[1] as a new $f_0(1800)$ resonance since in the $K\overline{K}$ decay channel the mass of $f_0(1800)$ would be very far from the $K\overline{K}$ threshold and the peak should be clearly observable, with no ambiguities about its interpretation. Yet, in the experiment studying J/Ψ decay into $\gamma K\overline{K}$, clear peaks are seen for the $f_0(1500)$ and $f_0(1710)$ but no trace is seen of any peak around 1800 MeV.[9] Similarly, MARK III[10] reports a clear signal for the $f_0(1710)$ in the $K\overline{K}$ spectra but no signal around 1800 MeV.

The suppressed decay of $f_0(1790)$ of Ref. 2 to $K\bar{K}$, however, can be understood in terms of its $\pi\pi f_0(980)$ resonance nature found in Ref. 4, where we first studied three pseudoscalar systems with total strangeness zero by solving the Faddeev equations in a coupled channel approach. For this, we obtained that input two-body amplitudes following Refs. 11,12 where dynamical generation of the light scalar mesons was found. Thus, our two-body amplitudes also contain this information. To be precise, the isoscalar $K\overline{K}$ and $\pi\pi$ t-matrices dynamically generate the resonances $f_0(980)$ and

$\sigma(600)$, while the system composed of the channels $K\overline{K}$ and $\pi\eta$ in isospin 1 gives rise to the $a_0(980)$ state. In the strangeness $+1$ $K\pi$ and $K\eta$ systems the $\kappa(850)$ is formed. As a result of solving the Faddeev equations with these inputs we find a resonance which can be associated with the $\pi(1300)$. Next, we use this three pseudoscalar amplitude and solve the Faddeev equations, once again, for the $\pi\pi f_0(980)$ system and we end up finding a resonance whose quantum numbers and mass match well with the $f_0(1790)$ resonance. This structure of $f_0(1790)$ is such that it would decay to two/four pions or $\pi\pi K\overline{K}$ channels but not to $K\overline{K}$,[3,4] as found in Ref. 2.

With this we can summarize the present manuscript by mentioning that we provide evidence for the existence of a new scalar, isoscalar resonance $f_0(1790)$ which is distinct to the known $f_0(1710)$. We present an explanation for the suppressed decay of this resonance to $K\overline{K}$. We also show that the BES data on the $\phi\omega$ spectrum can be explained in terms of $f_0(1710)$ and thus a new $f_0(1800)$ is not required. Thus there are two f_0 states in the 1700-1800 MeV.

References

1. M. Ablikim *et al.* [BES Collaboration], *Phys. Rev. Lett.* **96**, 162002 (2006); G. Huang [BESIII Collaboration], arXiv:1209.4813 [hep-ex].
2. M. Ablikim *et al.* [BES Collaboration], *Phys. Lett. B* **607**, 243-253 (2005).
3. K. P. Khemchandani, A. M. Torres, M. Nielsen, F. S. Navarra, D. Jido, A. Hosaka and E. Oset, arXiv:1311.4697 [hep-ph].
4. A. Martinez Torres, K. P. Khemchandani, D. Jido and A. Hosaka, *Phys. Rev. D* **84**, 074027 (2011); A. Martinez Torres, K. P. Khemchandani, D. Jido, Y. Kanada-Enyo and A. Hosaka, *Few Body Syst.* **54**, 333 (2013).
5. A. Martinez Torres, K. P. Khemchandani, F. S. Navarra, M. Nielsen and E. Oset, Phys. Lett. B **719**, 388 (2013).
6. L. S. Geng and E. Oset, *Phys. Rev. D* **79**, 074009 (2009).
7. Q. Zhao and B. -S. Zou, *Phys. Rev. D* **74**, 114025 (2006).
8. L. S. Geng, F. K. Guo, C. Hanhart, R. Molina, E. Oset and B. S. Zou, *Eur. Phys. J. A* **44**, 305 (2010).
9. J. Z. Bai *et al.* [BES Collaboration], *Phys. Rev. D* **68**, 052003 (2003).
10. W. Dunwoodie [MARK-III Collaboration], *AIP Conf. Proc.* **432**, 753 (1998).
11. J. A. Oller, E. Oset, *Nucl. Phys. A* **620** , 438-456 (1997).
12. J. A. Oller, E. Oset and J. R. Pelaez, *Phys. Rev. D* **59**, 074001 (1999).D 60, 099906].

Ab initio no core full configuration approach for light nuclei

Youngman Kim* and Ik Jae Shin†

Rare Isotope Science Project, Institute for Basic Science, Daejeon 305-811, Korea
* E-mail: ykim@ibs.re.kr
† E-mail: geniean@ibs.re.kr

Pieter Maris‡ and James P. Vary§

Department of Physics and Astronomy, Iowa State University, Ames, IA 50011, USA
‡ E-mail: pmaris@iastate.edu
§ E-mail: jvary@iastate.edu

Christian Forssén¶ and Jimmy Rotureau**

Department of Fundamental Physics, Chalmers University of Technology, SE-412 96
Göteborg, Sweden
¶ E-mail: christian.forssen@chalmers.se
** E-mail: rotureau@chalmers.se

Comprehensive understanding of the structure and reactions of light nuclei poses theoretical and computational challenges. Still, a number of *ab initio* approaches have been developed to calculate the properties of atomic nuclei using fundamental interactions among nucleons. Among them, we work with the *ab initio* no core full configuration (NCFC) method and *ab initio* no core Gamow Shell Model (GSM). We first review these approaches and present some recent results.

Keywords: Nuclear structure; *ab initio* approach.

1. Introduction

Nuclei are complicated quantum many-body systems and offer a solid testing ground for our knowledge of the strong interaction in the non-perturbative regime. It is a formidable task to get a firm grasp of how stable (and unstable) nuclei emerge from protons and neutrons whose interactions are dominated by the strong interaction. With the rapid growth of available high performance supercomputers, several *ab initio* approaches have been developed to study nuclear structures and reactions based on fundamental nuclear interactions. Robust and reliable results from *ab initio* methods may provide a clue to the role of fundamental degrees of freedom such as quarks in nuclei.

In this work, we study the properties of ^{6}Li using the *ab initio* no core full configuration (NCFC)[1] and no core Gamow Shell Model (GSM)[2] approaches with two different nucleon-nucleon (NN) interactions; the inverse-scattering interaction JISP16 and the new NNLO$_{opt}$ potential[3] from chiral effective field theory (chEFT). We first review these approaches and present some of our recent results.[4]

2. *Ab initio* No Core Full Configuration Approach

We start with the configuration interaction (CI) method on which the *ab initio* no core full configuration (NCFC) method is based. In short, the CI method is one of the post Hartree-Fock methods for solving the Schrödinger equation using a matrix formulation. The A-body Schrödinger equation is

$$\hat{H}\Psi(\mathbf{r}_1, \mathbf{r}_2, \ldots, \mathbf{r}_A) = E\Psi(\mathbf{r}_1, \mathbf{r}_2, \ldots, \mathbf{r}_A), \tag{1}$$

where the Hamiltonian $\hat{H}$ contains kinetic energy and interaction terms. In contrast to the Hartree-Fock method, where the A-body wave function is approximated by a single Slater determinant, the A-body wave function in the CI method is given by a linear combination of Slater determinants Φ_i:

$$\Psi(\mathbf{r}_1, \mathbf{r}_2, \ldots, \mathbf{r}_A) = \sum_{i=0}^{k} c_i \Phi_i(\mathbf{r}_1, \mathbf{r}_2, \ldots, \mathbf{r}_A). \tag{2}$$

To obtain the exact A-body wave function one has to consider infinite number of configurations, $k = \infty$, in practice, however, the sum must be limited to a finite number of configurations. The Slater determinant is the antisymmetrized product of single particle wave functions $\phi_\alpha(\mathbf{r})$, where α denotes the quantum numbers of single particle states. A traditional choice for the single particle basis is that of harmonic oscillator. Now, the matrix elements of the Hamiltonian is given by $H_{ij} = \langle \Phi_i | \hat{H} | \Phi_j \rangle$. For large and sparse matrices, the Lanczos method[5] has been widely used to find the extreme eigenvalues. This method is implemented in MFDn,[6–8] a hybrid MPI/OpenMP CI code for *ab initio* nuclear structure calculations.

Now we move on to the NCFC approach. This method is a version of the *ab initio* no core shell model (NCSM) with a few important characteristics that will be outlined below. The NCSM treats all nucleons in a nucleus as active and dynamical degrees of freedom. There is no postulated closed, inert core of nucleons in the nucleus. In the *ab initio* NCSM we start with the intrinsic Hamiltonian of A nucleons

$$H_A = \frac{1}{A} \sum_{i<j} \frac{(\mathbf{p}_i - \mathbf{p}_j)^2}{2m} + \sum_{i<j} V_{\mathrm{NN},ij} + \sum_{i<j<k} V_{\mathrm{NNN},ijk} \tag{3}$$

and add the harmonic oscillator (HO) center of mass Hamiltonian. Here, m is the nucleon mass, and V_{NN} (V_{NNN}) is a two-nucleon (three-nucleon) interaction. In the NCSM, the HO basis is employed. Due to the strong short-range correlations of nucleons in a nucleus, a large model space is required to achieve convergence. This infinite (or very large) model space problem might be overcome by the use of effective interactions rather than bare ones. For more on the NCSM, we refer to a recent review article.[9]

Features of the NCFC approach are: (1) the use of interactions defined for an infinite Hilbert space, (2) extrapolating to the continuum limit (infinite matrix limit), and (3) uncertainty estimation for the extrapolation.

Next, we discuss the interactions adopted in the current work. In the present study, we are using the JISP16 phenomenological and NNLO$_{\mathrm{opt}}$ chiral NN potentials. JISP (J-matrix Inverse Scattering Potential) type interactions[10–12] are constructed in the framework of the J-matrix version of inverse scattering theory. The matrix elements of the NN potential are calculated in the oscillator basis for each partial wave to reproduce experimental NN scattering data and deuteron properties without three-nucleon interactions. The JISP16 potential is obtained to fit the experimental data for light nuclei up to $A = 16$. A promising approach to construct and understand the nuclear force from first principles is chEFT.[13] An important and up-to-date optimization of the chiral Next-to-Next-to-Leading Order (NNLO) potential was performed using POUNDERS (Practical Optimization Using No Derivatives for Squares), to obtain the potential we label as NNLO$_{\mathrm{opt}}$.[3] The new chiral NNLO$_{\mathrm{opt}}$ yields $\chi^2 \approx 1$ per degree of freedom for laboratory energies roughly less than 125 MeV. It is also observed that the effects of three-nucleon interactions on the properties of light nuclei with $A = 3, 4$ are smaller than previously available parameterizations of chiral nuclear forces.

We now address the extrapolation to infinite matrix limit. We work with the N_{max} truncation scheme, where N_{max} is the basis truncation parameter. In this scheme, we consider all possible configurations with N_{max} excitations above the unperturbed ground state: $\sum N_i \leq N_0 + N_{\mathrm{max}}$. Here, N_0 is the total number of HO quanta for the ground state configuration and N_i is the number of quanta for each state. To take the infinite matrix limit, several extrapolation methods have been developed.[1,14–17]

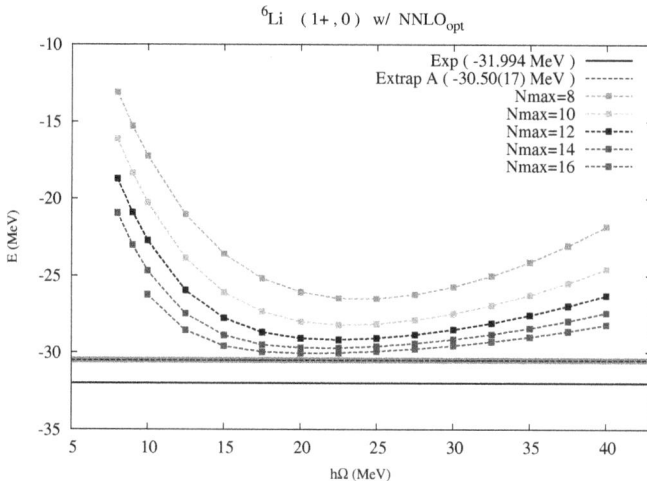

Fig. 1. Ground state energy of ^{6}Li calculated with NNLO$_{opt}$ as a function of the size of HO basis N_{max} and the result with the extrapolation A. The shaded area around the extrapolation A result indicates our estimated uncertainty of 170 keV.

Finally, we show a few results from our NCFC study.[4] In Fig. 1, we show the convergence of the ground state energy of ^{6}Li with the extrapolation A,[1] while excitation energies are presented in Fig. 2. The results are obtained from computations in model spaces up to $N_{max} = 16$ (matrix dimension 8×10^8). For a previous study on the ^{6}Li in the *ab initio* NCFC method, we refer to the work by Cockrell *et al.*,[18] where lithium isotopes, ^{6}Li, ^{7}Li, and ^{8}Li, are studied with the JISP16 interaction.

3. The *ab initio* Gamow Shell Model

As one approaches the particle emission thresholds, it becomes increasingly important to describe correctly the coupling to the continuum of decays and scattering channels. The recently developed complex-energy Gamow Shell Model (GSM)[19] has proven to be a reliable tool in the description of nuclei, where continuum effects cannot be neglected. In the GSM, the many-body basis is constructed from a single-particle Berggren ensemble[20] which includes bound, resonant and complex-continuum states. For practical calculations, the set of continuum states is discretized. As in any Shell Model calculation the dimension of the Hamiltonian matrix grows rapidly with the number of single-particle states and the number of nucleons. In addition, the Hamiltonian matrix in our rigged Hilbert space is non-

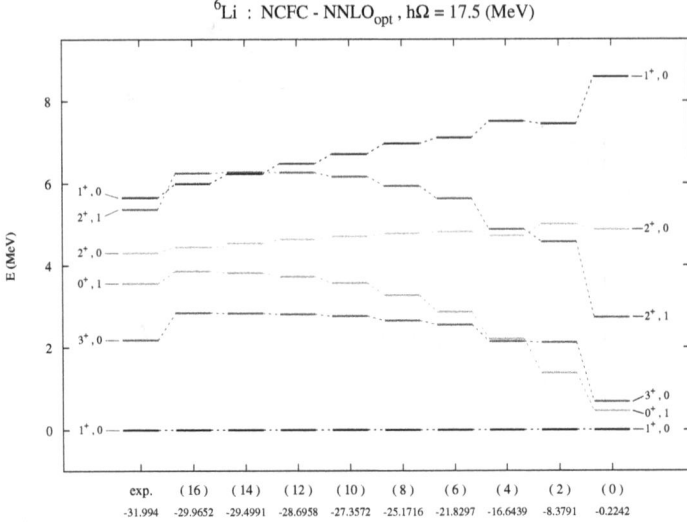

^{6}Li : NCFC - NNLO$_{opt}$, $\hbar\Omega = 17.5$ (MeV)

Fig. 2. Excitation energies of ^{6}Li calculated with NNLO$_{opt}$ and experimental data. The low-lying positive parity states are shown as a function of N_{max} truncation (indicated in parenthesis below each column). The ground state eigenvalue (in MeV) is also listed for each N_{max}.

Hermitian (complex symmetric). Hence, advanced numerical methods that can handle large non-Hermitian matrices must be used. In the context of the GSM, it has been shown that the Density Matrix Renormalization Group (DMRG) is an efficient way to compute the low-lying spectrum of the Hamiltonian at a low computational cost.[21]

Let us consider the application of the J-scheme DMRG in the context of the GSM (GSM+DMRG). The objective is to calculate an eigenstate $|J^\pi\rangle$ of the GSM Hamiltonian $\hat{H}$ with angular momentum J and parity π. As $|J^\pi\rangle$ is a many-body pole of the scattering matrix of $\hat{H}$, the contribution from non-resonant scattering shells along the continuum contour L^+ to the many-body wave function is usually smaller than the contribution from the resonant orbits.[19] Based on this observation, the following separation is usually performed:[21] the many-body states constructed from the single-particle poles form a subspace A (the so-called 'reference subspace'), and the remaining states containing contributions from non-resonant shells form a complement subspace B.

One begins by constructing states $|k\rangle_A$ forming the reference subspace A. All possible matrix elements of suboperators of the GSM Hamiltonian $\hat{H}$ acting in A, expressed in the second quantization form, are then calculated and stored and the GSM Hamiltonian is diagonalized in the reference space to provide the zeroth-order approximation $|\Psi_J\rangle^{(0)}$ to $|J^\pi\rangle$. This vector, called 'reference state', plays an important role in the GSM+DMRG truncation algorithm. The scattering shells (lj), belonging to the discretized contour L^+, are then gradually added to the reference subspace to create the subspace B. This first stage of the GSM+DMRG procedure is referred to as the warm-up phase. For each new shell that is added, all possible many-body states denoted as $|i\rangle_B$ are constructed and matrix elements of suboperators of the GSM Hamiltonian acting on $|i\rangle_B$ are computed. By coupling states in A with the states $|i\rangle_B$, one constructs the set of states of a given J^π. This ensemble serves as a basis in which the GSM Hamiltonian is diagonalized. The target state $|\Psi_J\rangle$ is selected among the eigenstates of $\hat{H}$ as the one having the largest overlap with the reference vector $|\Psi_J\rangle^{(0)}$. Then, the desired truncation is performed in B by introducing the reduced density matrix, constructed by summing over the reference subspace A. The GSM density matrix being complex-symmetric, the truncation is done by keeping the eigenstates α_B (the 'optimized' states) with the largest nonzero moduli of eigenvalue w_α.[21]

The warm-up phase is followed by the so-called sweeping phase, in which, starting from the last scattering shell $(lj)_{\text{last}}$, the procedure continues in the reverse direction (the 'sweep-down' phase) until the first scattering shell is reached. The procedure is then reversed and a sweep in the upward direction (the 'sweep up' phase) begins. The sweeping sequences continue until convergence for target eigenvalue is achieved.

A no core GSM+DMRG approach was recently developed[2] to be used for *ab initio* studies of light nuclei using realistic interactions. Here we show an application of the DMRG method for the $J^\pi = 1^+$ ground state in ^{6}Li. Since this state is well bound, the effects of the coupling to the continuum states are negligible. Nevertheless, for the purpose of illustration, we show results using the DMRG technique in a model space containing only HO shells. The model space includes proton and neutron shells with energy up to 10 $\hbar\omega$ that is, we include s-shells up to the $5s_{1/2}$, p-shells up to $4p_{1/2;3/2}$ and d-shells up to $4d_{3/2;5/2}$. For this calculation which serves as an illustration of the method, we are not including shells with higher l-values. In Fig. 3 we show results obtained by keeping the eigenstates of the density matrix such that $\epsilon = 1 - \sum_\alpha w_\alpha \leq 5 \times 10^{-6}$. Results are shown starting from the middle of the warm-up phase until the end of the second sweep. The relatively small difference between the lowest and highest energy during the second sweep (~ 360 keV) could be further decreased by keeping

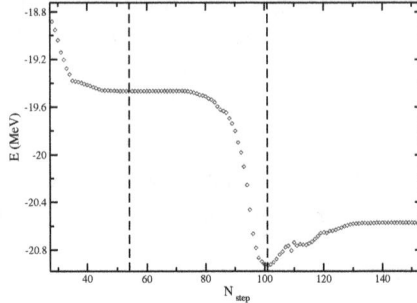

Fig. 3. Iterative process of the DMRG approach for $\epsilon = 5 \times 10^{-6}$ and including only waves up to $l = 2$ (s,p,d). Results are shown starting from the middle of the warm-up phase, and the two vertical dashed lines show respectively the beginning of the first and second sweeping phase.

more states.[21] The dimension of the total model space in the J-scheme is 141,762,900 whereas the largest DMRG matrix to be diagonalized has a dimension equal to 68,386.

4. Summary

We briefly introduced the *ab initio* NCSM, NCFC method, and *ab initio* GSM approach. To study the properties of ^{6}Li, we employed the JISP16 realistic nucleon-nucleon potential and chiral NNLO$_{\rm opt}$ interaction. We showed some of our recent results in Figs. 1–3. From Figs. 1 and 2, we conclude that sufficient convergence is achieved in our study.

Acknowledgments

The work of YK and IJShin was supported by the Rare Isotope Science Project of Institute for Basic Science funded by Ministry of Science, ICT and Future Planning and National Research Foundation of Korea (2013M7A1A1075766). This work was supported in part by the U.S. National Science Foundation under Grant No. PHY-0904782 and the U.S. Department of Energy under Grant Nos. DE-FG02-87ER40371 and DESC0008485 (SciDAC-3/NUCLEI). The research leading to these results has received funding from the European Research Council under the European Community's Seventh Framework Programme (FP7/2007-2013)/ERC grant agreement no. 240603. A portion of the computational resources were provided by the National Energy Research Scientific Computing Center (NERSC), which is supported by the U.S. DOE Office of Science under Contract No. DE-AC02-05CH11231. Computational resources were also provided by the Supercomputing Center/Korea Institute of Science and Technology Information including technical support (KSC-2012-C3-054).

References

1. P. Maris, J. P. Vary and A. M. Shirokov, *Phys. Rev. C* **79** (2009) 014308.
2. G. Papadimitriou, J. Rotureau, N. Michel, M. Płoszajczak and B. R. Barrett, *Phys. Rev. C* **88** (2013) 044318.
3. A. Ekstrm, G. Baardsen, C. Forssn, G. Hagen, M. Hjorth-Jensen, G. R. Jansen, R. Machleidt and W. Nazarewicz *et al.*, *Phys. Rev. Lett.* **110** (2013) 192502.
4. C. Forssén, Y. Kim, P. Maris, J. Rotureau, I. J. Shin and J. P. Vary, *in preparation*.
5. B. N. Parlett, *The Symmetric Eigenvalue Problem*, Prentice-Hall, 1980.
6. P. Sternberg, E. G. Ng, C. Yang, P. Maris, J. P. Vary, M. Sosonkina and H. V. Le, *Accelerating configuration interaction calculations for nuclear structure*, in *Proc. of the 2008 ACM/IEEE conf. on Supercomputing*, IEEE Press, Piscataway, NJ, p. 15:1 (2008).
7. P. Maris, M. Sosonkina, J. P. Vary, E. G. Ng and C. Yang, Proc. Comput. Sci. **1**, 97 (2010).
8. H. M. Aktulga, C. Yang, E. G. Ng, P. Maris and J. P. Vary, *Improving the scalability of symmetric iterative Eigensolver for multi-core platforms*, Concurrency Computat.: Pract. Exper. DOI: 10.1002/cpe.3129 (2013, in press).
9. B. R. Barrett, P. Navratil and J. P. Vary, *Prog. Part. Nucl. Phys.* **69** (2013) 131.
10. A. M. Shirokov, A. I. Mazur, S. A. Zaytsev, J. P. Vary and T. A. Weber, *Phys. Rev. C* **70** (2004) 044005.
11. A. M. Shirokov, J. P. Vary, A. I. Mazur, S. A. Zaytsev and T. A. Weber, *Phys. Lett. B* **621** (2005) 96.
12. A. M. Shirokov, J. P. Vary, A. I. Mazur and T. A. Weber, *Phys. Lett. B* **644** (2007) 33.
13. R. Machleidt and D. R. Entem, *Phys. Rept.* **503** (2011) 1.
14. P. Maris, A. M. Shirokov and J. P. Vary, *Phys. Rev. C* **81** (2010) 021301
15. S. A. Coon, M. I. Avetian, M. K. G. Kruse, U. van Kolck, P. Maris and J. P. Vary, *Phys. Rev. C* **86** (2012) 054002.
16. R. J. Furnstahl, G. Hagen and T. Papenbrock, *Phys. Rev. C* **86** (2012) 031301
17. E. D. Jurgenson, P. Maris, R. J. Furnstahl, P. Navratil, W. E. Ormand and J. P. Vary, *Phys. Rev. C* **87** (2013) 054312.
18. C. Cockrell, J. P. Vary and P. Maris, *Phys. Rev. C* **86** (2012) 034325.
19. N. Michel, W. Nazarewicz, M. Płoszajczak and T. Vertse, *J. Phys. G* **36** (2009) 013101.
20. T. Berggren, *Nucl. Phys. A* **109** (1968) 265.
21. J. Rotureau, N. Michel, W. Nazarewicz, M. Płoszajczak and J. Dukelsky, *Phys. Rev. C* **79** (2009) 014304.

Understanding nuclear shape phase transitions within SD-pair shell model

Lei Li

School of Physics, Nankai University, Tianjin, 300071, P. R. China

Yu Zhang

Department of Physics, Liaoning Normal University, Dalian 116029, P. R. China

Xiaoqing Yuan

School of Physics, Nankai University, Tianjin, 300071, P. R. China

Jiangdan Li

School of Physics, Nankai University, Tianjin, 300071, P. R. China

Yanan Luo

School of Physics, Nankai University, Tianjin, 300071, P. R. China
E-mail: luoya@nankai.edu.cn

Feng Pan

Department of Physics, Liaoning Normal University, Dalian 116029, P. R. China
Department of Physics and Astronomy, Louisiana State University,
Baton Rouge, LA 70803, USA

Jerry P. Draayer

Department of Physics and Astronomy, Louisiana State University,
Baton Rouge, LA 70803, USA

The effect of strength of each interaction on the nuclear shape phase and their transitional patterns are studied in the SD-pair shell model with a Hamiltonian composed of the single-particle energy term, monopole-pairing, quadrupole-pairing and quadrupole-quadrupole interaction for identical nucleon system. It is shown that with quandrupole-pairing interaction and quadrupole-quadrupole interaction strengths set to be 0, the nuclear phase transition from single-particle motion to collective motion can be produced by changing the monopole pairing interaction. With fixed monopole-pairing interaction and quadrupole-pairing interactional strength set to be 0, the like-vibration-rotation shape phase transitional pattern can be produced by changing the quadrupole-quadrupole interaction from 0 to $0.1\mathrm{MeV}/r_0^4$.

1. Introduction

Nuclei, as a mesoscopic system, have been found to possess interesting geometric shapes, such as spherical vibrational $(U(5))$, axially deformed $(SU(3))$, and γ-soft $(O(6))$, which are usually described in terms of the Casten triangle in the interacting boson model (IBM).[1] The search for signatures of transitions among various shapes (phases) of atomic nuclei is an interesting subject in nuclear structure theory and has been studied extensively in recent years. An understanding of such shape phase transitions may provide insight into quantum phase transitions in other mesoscopic systems.[2] Theoretical study of shape phase transitions and critical point symmetries in nuclei has mainly been carried out[2–21] in the IBM.[2] The IBM is a phenomenological model of nuclear structure which has a deep connection with the microscopic shell model.[22,23] A long-standing significant question is then to identify directly the shape phase structure, especially, the shape phase transition, in fermion space or at nucleon level. Recently there have been studies on nuclear shape phase transitions and their critical point symmetries in the framework of shell model,[24–30] density functional approach[31] and relativistic mean field approach.[32]

The investigations on nuclear shape phase transition and critical point symmetry for identical nucleon system have also been carried out with fermionic degrees of freedom in Refs. 33–38.

In Ref. 39, a correspondence between the strength of each of the interactions and the nuclear shape phases is obtained with the Dyson boson mapping approach, in which a shell model Hamiltonian with monopole-pair, quadrupole-pair and quadrupole-quadrupole interactions between nucleons were used. The results show that increasing the quadrupole-pair interaction strength can induce the vibrational to the axially prolate rotational shape phase transition and enhancing the quadrupole-quadrupole interaction can drive the phase transition from the axially oblate rotational to the axially prolate rotational, with the γ-soft rotational being the critical point.

SD-pair shell model(SDPSM) is built up from SD fermion pairs, and the Hamiltonian is diaganolized in the fermion space directly.[41] Our previous work in the SDPSM show that the vibrational, rotational, triaxial and γ-soft spectra can be well reproduced[42,43] similar to the $U(5)$, $SU(3)$, $SU^*(3)$ and $SO(6)$ limiting spectra in the IBM. The nuclear shape phase transitional patterns and properties of the critical point symmetry can also be produced within the framework of the SDPSM with fermionic degrees of freedom.[44,45] It is interesting to see if the similar conclusion as in Ref. 39 can be obtained in the SDPSM. This is the main objective of this paper.

2. Model

As in Ref. 39, we take a general shell model Hamiltonian[28,30] to study the dependence of the shape phases on each of the interactions, which is a combination of the single particle energy, monopole pairing, quadrupole-pairing and quadrupole-quadrupole interaction with

$$H = H_0 - G_0 \mathcal{S}^\dagger \mathcal{S} - G_2 \mathcal{P}^\dagger \mathcal{P} - \kappa Q^{(2)} \cdot Q^{(2)} \tag{1}$$

$$H_0 = \sum_a \epsilon_a n_a$$

$$\mathcal{S}^\dagger = \sum_a \frac{\widehat{j_a}}{2} \left(C_a^\dagger \times C_a^\dagger \right)$$

$$\mathcal{P}^\dagger = \sum_{ab} q(ab2) \left(C_a^\dagger \times C_b^\dagger \right)^2$$

$$Q_\mu^{(2)} = \sqrt{\frac{16\pi}{5}} \sum_{i=1}^{n} r_i^2 Y_{2\mu}(\theta_i \phi_i)$$

where a denote all quantum number necessary to specify a state $[a \equiv (nlj)]$. ε_a and n_a are the single-particle energy and number operator of state a, $\widehat{j_a} = \sqrt{2j_a + 1}$ respectively. G_0, G_2 and κ is the monopole-pairing, quadrupole-pairing and quadrupole-quadrupole interaction strength, respectively. $Q_\mu^{(2)}$ is the quadrupole operator which in second quantized form is given by

$$Q_\mu^{(2)} = \sum_{cd} q(cd2) P_\mu^2(cd),$$

$$q(cd2) = (-)^{c-\frac{1}{2}} \frac{\widehat{c}\widehat{d}}{\sqrt{20\pi}} C_{c\frac{1}{2},d-\frac{1}{2}}^{2\ 0} \Delta_{cd2} \langle Nl_c | r^2 | Nl_d \rangle,$$

$$\Delta_{cd2} = \frac{1}{2} \left[1 + (-)^{l_c + l_d + 2} \right],$$

$$P_\mu^t(cd) = \left(C_c^\dagger \times \tilde{C}_d \right)_\mu^t,$$

where in this expression $\widehat{l} \equiv \sqrt{2l+1}$, $C_{c\frac{1}{2},d-\frac{1}{2}}^{2\ 0}$ is an ordinary $SU(2)$ Clebsch-Gordan (CG) coefficient, and N is the principal quantum number of the harmonic oscillator wave function with energy eigenvalue $(N + 3/2)\hbar\omega_0$. The matrix elements for r^2 are

$$\langle Nl_c | r^2 | Nl_d \rangle = \begin{cases} (N + 3/2)r_0^2, & l_c = l_d, \\ \varphi[(N + l_d + 2 \pm 1)(N - l_d + 1 \mp 1)]^{1/2}r_0^2, & l_c = l_d \pm 2, \end{cases} \tag{2}$$

where the phase factor φ can be taken either as -1 or $+1$, and $r_0^2 = \hbar/M_N\omega_0 = 1.012A^{1/3}fm^2$, M_N is the mass of a nucleon, and ω_0 is frequency of the harmonic oscillator.

The $E2$ transition operator is simply

$$T(E2) = e_{\text{eff}}Q^{(2)}, \tag{3}$$

where e_{eff} is the effective charge.

The collective pairs $A_\mu^{r\dagger}$ with $r = 0$, 2 and the angular momentum projection μ built from many non-collective pairs $\left(C_a^\dagger \times C_b^\dagger\right)_\mu^r$ in the single-particle levels a and b are

$$A_\mu^{r\dagger} = \sum_{ab} y(abr) \left(C_a^\dagger \times C_b^\dagger\right)_\mu^r, \tag{4}$$

$$y(abr) = -\theta(abr)y(bar), \quad \theta(abr) = (-)^{j_a+j_b+r},$$

where $y(abr)$ are structure coefficients. In this paper, as an approximation, the S-pair structure coefficients are determined as $y(aa0) = \sqrt{2j_a + 1}\frac{v_a}{u_a}$, where v_a and u_a are the occupied and unoccupied amplitudes for orbit a obtained by solving the associated BCS equation. The D pair is obtained by using the commutator,

$$D^\dagger = \tfrac{1}{2}[Q^2, S^\dagger] = \sum_{ab} y(ab2) \left(C_a^\dagger \times C_b^\dagger\right)^2. \tag{5}$$

To see if the SDPSM can produce the similar results as in Ref. 39, the same major shell and single particle energy levels as in Ref. 39 are used, which is 1.3, 2.8, 0, 0.8 and 2.5 for $j = 1/2$, $3/2$, $5/2$, $2/7$ and $11/2$, respectively. The single-particle wave functions are taken to be the harmonic oscillators with oscillation constant $b_2 = 1.0A^{1/3}fm^2$. In addition, for simplicity, protons and neutrons are not differentiated, and the single-particle orbits energies are set to be invariant in the whole calculations. The effective charge is set to be 1.0 in the electric properties calculations.

Quantities of interest are the normalized low-lying levels energies and the electric quadrupole transition rates. The low-lying levels energies are normalized to $E_{2_1^+}$, and the B(E2)s are normalized to $B(E2; 2_1^+ \to 0_1^+)$. For the convenience of expression, we denote the characteristic quantities as $R_{4_1} = E_{4_1^+}/E_{2_1^+}$, $R_{6_1} = E_{6_1^+}/E_{2_1^+}$, $R_{0_2} = E_{0_2^+}/E_{2_1^+}$, $R_{0_3} = E_{0_3^+}/E_{2_1^+}$, $R_{2_2} = E_{2_2^+}/E_{2_1^+}$, $B_{4_1 2_1} = \frac{B(E2;4_1^+ -> 2_1^+)}{B(E2;2_1^+ -> 0_1^+)}$, $B_{6_1 4_1} = \frac{B(E2;6_1^+ -> 4_1^+)}{B(E2;2_1^+ -> 0_1^+)}$, $B_{0_2 2_1} == \frac{B(E2;0_2^+ -> 2_1^+)}{B(E2;2_1^+ -> 0_1^+)}$, $B_{0_3 2_1} = \frac{B(E2;0_3^+ -> 2_1^+)}{B(E2;2_1^+ -> 0_1^+)}$ and $B_{2_2 2_1} = \frac{B(E2;2_2^+ -> 2_1^+)}{B(E2;2_1^+ -> 0_1^+)}$. It is known that these quantities can characterize the shape phase structure and transition, and some of them can even distinguish the first order from the second order phase transition (see, for example, Ref. 19).

3. Numerical Results

3.1. *The effect of the monopole pairing interaction*

As in Ref. 39, to study the effect of the monopole-pairing interaction on the collectivity of low-lying states, the quadrupole-pairing and quadrupole-quadrupole interactional strengths are set to be 0. Since the pair-structures are determined by the BCS theory in the SDPSM, if the monopole-pairing interaction strength is small, the distribution of the nucleons over the single-particle levels is not evenly, most of the nucleons will distribute in the low-lying single-particle levels. Therefore, to study the effect of the monopole-pairing strength on the low-lying states, the monopole-pairing strength G_0 is set to be as $0.05 \leq G_0 \leq 0.5$ MeV.

The calculated results of the dependence of the energy levels and the normalized ones on the monopole-pair interaction strength G_0 are illustrated in Fig. 1 and Fig. 2. The normalized $E2$ transition rates of the dependence of G_0 are display in Fig. 3.

From the three figures, it is seen that when $G_0 < 0.1$MeV, the position of the energy levels do not change with G_0 and the $B(E2)$ values are all close to be 0, i.e., the collectivity of low-lying states can not be produced for small G_0 values. The energy levels and $B(E2)$ values begin to increase with G_0 from $G_0 \sim 0.1$MeV. This analysis seems show that a new phase transition between single-particle motion and no-collective motion occurs in the SDPSM calculation.

One can also notice that the degenerate level structure of the vibrational states (U(5) symmetric states in the IBM)are reproduced for large G_0. For example, when $G_0 = 0.2$MeV, 0_2^+,2_2^+ and 4_1^+ are degenerate, 0_3^+, 2_3^+, 3_1^+, 4_2^+ and 6_1^+ are also degenerate, these are the typical feature of the vibrational limit. Therefore, a large G_0 favors a spherical phase.

Ref. 39 shows that as G_0 is very small, the calculated results show approximately the feature of the axially rotational phase. But from the above discussion one can see that the similar results can not be produced in the SDPSM.

3.2. *Effect of the quadrupole-pairing*

From Ref. 46 it is known that the quadrupole-pairing interaction is very important in producing the γ-soft spectrum for like-nucleon system. The

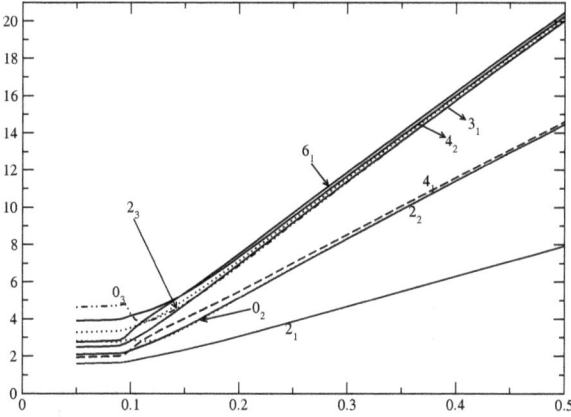

Fig. 1. Calculated result of the dependence of the low-lying levels energies on the monopole pairing interaction G_0 when $G_2 = \kappa = 0$.

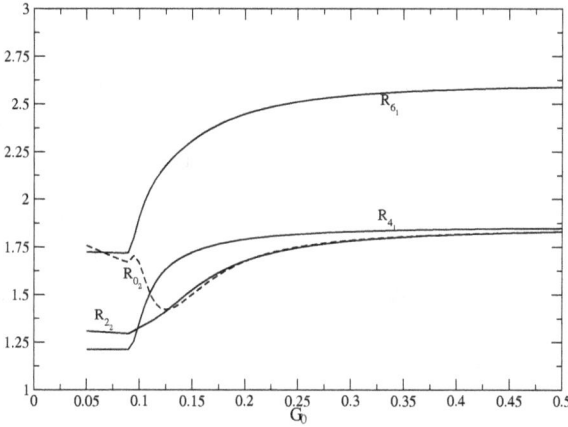

Fig. 2. Calculated results of the normalized energies on the monopole-pair interaction strength G_0 when $G_2 = \kappa = 0$.

effect of the quadrupole-pairing interaction on the collectivity of low-lying states were also be studied in Ref. 39. It is found that with $G_0 = 0.15$MeV and $\kappa = 0$, with the quadrupole-pair interaction strength $G_2 \in (0, 0.03)$, the system is in a vibrational phase approximately, and for $G_2 \in (0.06, 0.10)$, the axially rotational phase can be produced. It is interesting to see if the similar results about quadrupole-pairing interaction can be produced in the SDPSM.

To study the effect of the quadrupole pairing interaction on the collectivity of the low-lying states, we set $G_0=0.15$MeV and $\kappa = 0$ as in Ref. 39,

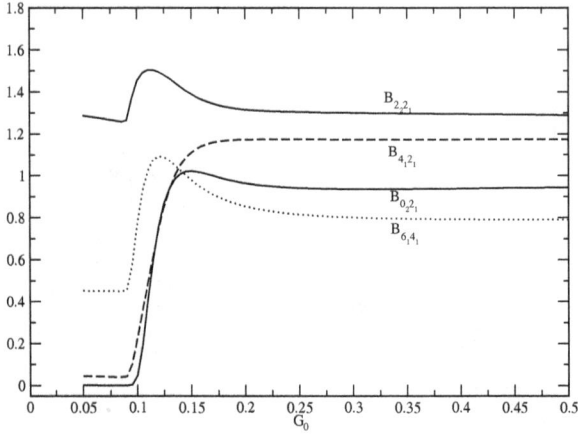

Fig. 3. Calculated result of the dependence of the normalized B(E2)s on the monopole-paring interaction strength G_0 when $G_2 = \kappa = 0$.

and $G_2 \in (0, 0.1)$. The results are given in Fig. 4, Fig. 5 and Fig. 6, 7 for energy, energy ratios and $B(E2)$ ratios, respectively. Since $G_0 = 0.15$ and $G_2 = \kappa = 0$ corresponds to the spherical phase, our calculation here shows in fact the effect of the quadrupole-pair interaction on the spherical phase as in Ref. 39.

The three figures show that the general behaviors in Ref. 39 can all be produced in the SDPSM. For example, all of the low-lying states decrease with G_2 till $G_2 = 0.05\text{MeV}/r_0^4$, then they all increase with G_2, i.e., the lowest point of all of the states is around $G_2 = 0.05\text{MeV}/r_0^4$. The relative energy ratios are given in Fig.5, from which one can see that except for R_{02} ,which has a lower point around $G_2 = 0.05\text{MeV}/r_0^4$, the similar behaviors of the other energy ratios as in Ref. 39 can be produced very well, i.e., they all increase with G_2 and reach the maximum point around $G_2 = 0.05\text{MeV}/r_0^4$. The behavior of 0_2^+ is opposite to the other states, namely, it decrease with G_2 when $G_2 \leq 0.05\text{MeV}$, after this point, it increases with G_2. But one can notice that the 0_2^+ and 0_3^+ states cross each other when $G_2 = 0.018\text{MeV}$, the same phenomena can also be found in the $SO(6)$ limiting cases in even-even Ba isotopes.[40] The relative B(E2) ratios are given with the results presented in Fig.6 and Fig.7. One can see that B_{22}, B_{64} and B_{42} decrease with G_2 slowly, from $G_2 = 0.05\text{MeV}/r_0^4$ on, they all increase with G_2, but for 0_2^+ and 0_3^+ states, one can see that the $B(E2)$ ratios exchange at $G_2 = 0.018\text{MeV}/r_0^4$. The behavior of the relative B(E2) ratios are constant with those in Ref. 39.

From above analysis one can see that considering the energy ratio and

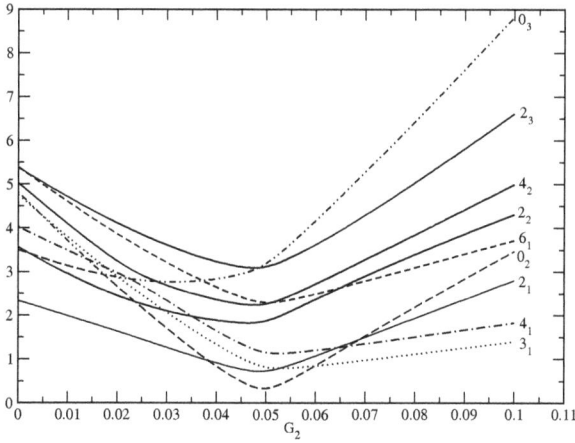

Fig. 4. Calculated result of the dependence of the low-lying levels energies on the quadrupole pairing interaction G_2 when $G_0 = 0.15\text{MeV}$ and $\kappa = 0$.

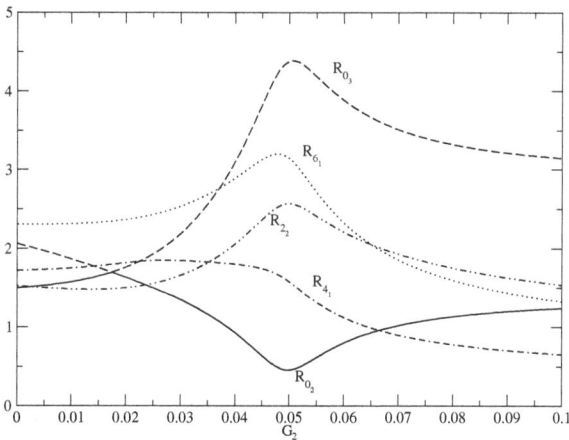

Fig. 5. Calculated result of the dependence of the energy ratios on the quadrupole pairing interaction G_2 when $G_0 = 0.15$ MeV and $\kappa = 0$.

$B(E2)$ values together, the properties of the 0_2^+ state in the SDPSM are also in consistent with that in Ref. 39. It is also seen that for $G_2 \in (0, 0.02)\text{MeV}/r_0^4$, the results are close to the vibrational phase, but for larger G_2 value, it is difficult to determine the shape phase of the nuclear system.

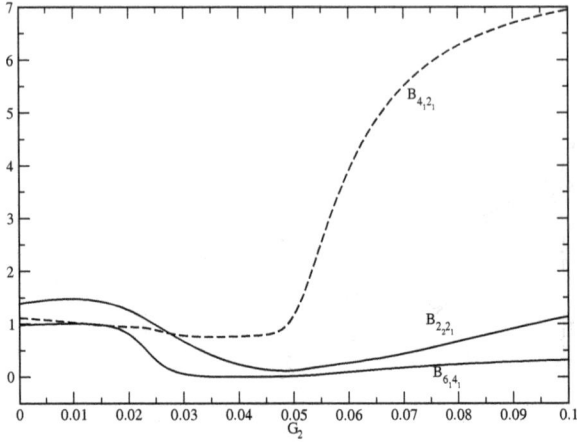

Fig. 6. Calculated result of the dependence of the $B(E2)$ ratios on the quadrupole pairing interaction G_2 when $G_0 = 0.15\text{MeV}$ and $\kappa = 0$.

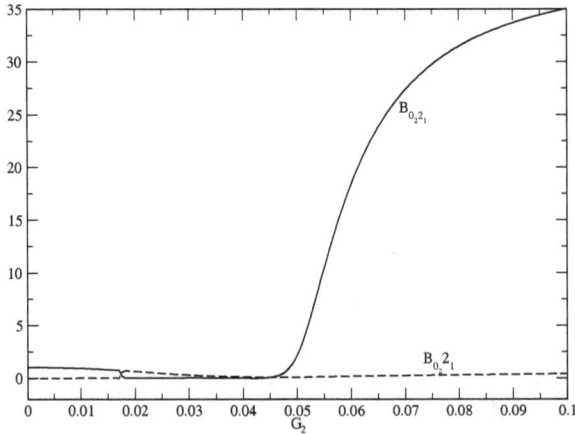

Fig. 7. Calculated result of the dependence of the $B(E2)$ ratios on the quadrupole pairing interaction G_2 when $G_0 = 0.15\text{MeV}$ and $\kappa = 0$.

3.3. *Effect of the quadrupole-quadrupole interactional strength*

It is known that the mono-pole pairing and quadrupole-quadrupole interactions are dominant shell-model interactions. As in Ref. 39, the effect of the quadrupole-quadrupole interactional strength are also studied in the SDPSM with $G_0=0.15\text{MeV}$, $G_2 = 0$ and $0 \leq \kappa \leq 0.1\text{MeV}/r_0^4$. Since

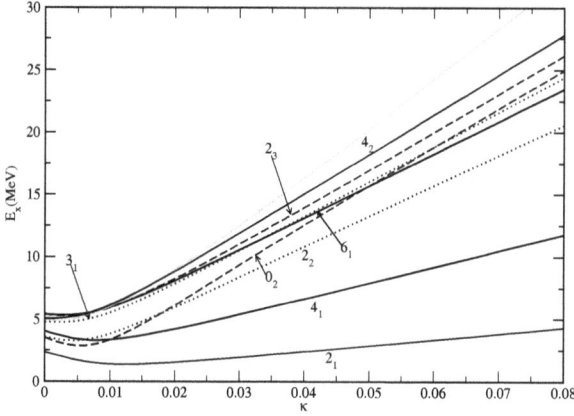

Fig. 8. Calculated result of the dependence of the low-lying levels energies on the quadrupole-quadrupole interaction κ when $G_0 = 0.15\text{MeV}$ and $G_2 = 0$.

$G_0 = 0.15\text{MeV}$ and $G_2 = 0$ corresponds to the spherical phase, our calculation here again shows in fact the effect of the quadrupole-quadrupole interaction on the spherical phase. The energies, energy ratios and $B(E2)$ ratios are given in Figs. 8, 9 and 10.

From Figs. 8 and Fig.9 one can notice that all the levels decrease with κ till $\kappa \sim 0.005\text{MeV}/r_0^4$, and the degenerate level structure of the vibrational states (U(5) symmetric states in the IBM), such as 0_2^+, 2_2^+ and 4_1^+ states, can be produced, while the level structure of the rotational states (SU(3) symmetric states in the IBM) can be reproduced for larger κ values.

To see whether the similar behaviors in the Ref. 39 can be produced, the relative $B(E2)$ ratios are given in Fig.10, it is seen that the $B(E2)$ ratios are close to the vibrational limit when κ is small, and $B_{2_2 2_1}$ and $B_{0_2 2_1}$ decrease with κ quickly. The results that $B_{4_1 2_1}$ and $B_{6_1 4_1}$ are much larger than $B_{2_2 2_1}$ and $B_{0_2 2_1}$ for large κ are the typical feature of the rotational limit.

4. Discussion and Summary

From above analysis one can see that for the case of the vibrational limit, R_{41} is smaller than the typical value of the vibrational limit 2.0, and for the case of the rotational limit, R_{42} is smaller than the typical value of the rotational limit 3.3. In this calculation, 50-82 shell are considered, and the neutrons and protons are treated as one kind of nucleons. Therefore, the pauli-blocking effect play an important role in producing the collectivity

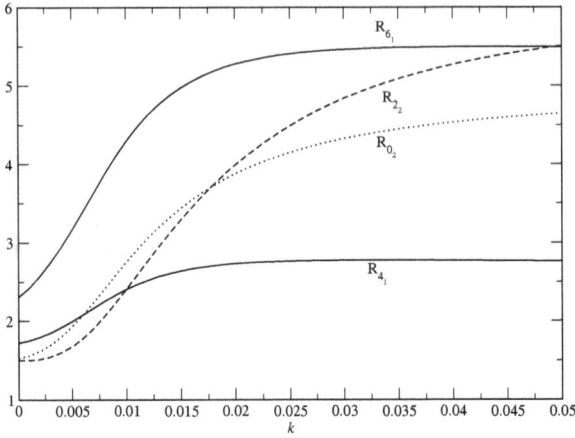

Fig. 9. Calculated result of the dependence of the energy ratios on the quadrupole-quadrupole κ when $G_0 = 0.15\text{MeV}$ and $G_2 = 0$.

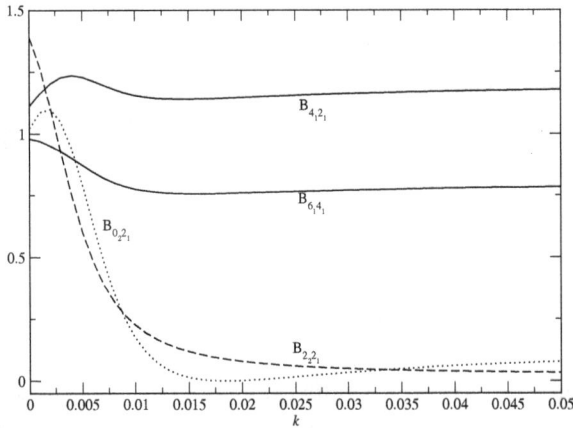

Fig. 10. Calculated result of the dependence of the $B(E2)$ ratios on the quadrupole-quadrupole interaction κ when $G_0 = 0.15\text{MeV}$ and $G_2 = 0$.

of the low-lying states. From Ref. 47 it is known that if a toy model like $j = 1/2, 3/2, 5/2, 7/2, 9/2, 11/2, 13/2, 15/2$ and $17/2$ is used for like nucleon system, the limiting cases in the IBM can be produced very well in the SDPSM.

In summary, the effect of the interactional strength on the nuclear shape phase transition patterns have been studied within the framework of the SD-pair shell model. The results show that by changing the monopole pairing interactional strength, the nuclear phase from single-particle motion to collective motion can be produced, i.e., if the monopole pairing interactional strength is small, the results are close to the single-particle motion in the SDPSM, from $G_0 = 0.1$MeV on, the collectivity of low-lying states can be produced. It is also shown that the shape phase transitional patterns of like-vibration-rotation in the identical nucleon system can be produced by changing the quadropole-qudrupole interactional strength in the SDPSM. This results also show that the results obtained in Ref. 39 about the validity of the boson mapping is reasonable if the general behavior of the vibration-rotation shape phase transitional patterns are considered.

One of the authors (Y.A. Luo) thanks Prof. Y.X. Liu, P. Ring and R. V. Jolos for their valuable suggestions. This work was supported by the Natural Science Foundation of China (11005056,11175078,11075080,11075081, 10935001,11075052) and he Natural Science Foundation for Talent Training in Basic Science (J1103208).

References

1. R. F. Casten, in: F. Iachello (Ed.), Interacting Bose-Fermi System, Plenum, New York, 1981.
2. F. Iachello and A. Arima, *The Interacting Boson Model*, Cambridge University Press, Cambridge New York, 1987.
3. J. N. Ginocchio and M. W. Kirson, Phys. Rev. Lett. **44**, 1744(1980).
4. D. H. Feng, R. Gilmore, and S. R. Deans, Phys. Rev. **23**, 1254(1981).
5. P. Van Isacker and J. Q. Chen, Phys. Rev. **C 24**, 684(1981).
6. F. Iachello, N. V. Zamfir and R. F. Casten, Phys. Rev. Lett. **81**, 1191(1998).
7. R. F. Casten and N. V. Zamfir, Phys. Rev. Lett. **85**, 3584 (2000).
8. F. Iachello, Phys. Rev. Lett. **87**, 052502(2001).
9. J. Jolie, P. Cejnar, R. C. casten, S. Heinze, A. Linnemann and V. Werner, Phys. Rev. lett. **89**, 182502(2002).
10. J. Jolie, R. C. casten, P. von Brentano and V. Werner, Phys. Rev. lett. **87**, 162501(2001).
11. D. Warner, Nature **420**, 614(2002).
12. F. Iachello and N. V. Zamfir, Phys. Rev. Lett. **92**, 212501(2004).
13. D. J. Rowe, Phys. Rev. Lett. **93**, 122502(2004).
14. D. J. Rowe, P. S. Turner and G. Rosenstell, Phys. Rev. Lett. **93**, 232502(2004).
15. P. Cejnar, S. Heinze and J. Dobeš, Phys. Rev. **C 71**, 011304(R)(2005).
16. Y. X. Liu, L. Z. Mu and H. Wei, Phys. Lett. **B 633**, 49(2006); Y. Zhao, Y. Liu, L. Z. Mu and Y. X. Liu, Int. J. Mod. Phys. **E 15**, 1711(2006).
17. F. Pan, J. P. Draayer, and Y. A. Luo, Phys. Lett. **B 576**, 297(2003).
18. A. leviatan, Phys. Rev. Lett. **77**, 818(1996); **98**, 242502(2007); A. leviatan and P. Van Isacker, Phys. Rev. Lett. **89**, 222501(2002).
19. Y. Zhang, Z. F. Hou and Y. X. Liu, Phys. Rev. **C 76**, 011305(R)(2007).

20. J. M. Arias, J. E. García-Ramos and J. Dukelsky, Phys. Rev. Lett. **93**, 212501(2004) and the reference cited in this paper.

21. M. A. Caprio and F. Iachello, Phys. Rev. Lett. **93**, 242502(2004).

22. T. Otsuka, A. Arima, F. Iachello, Nucl. Phys. A 309 (1978) 1; T. Otsuka, A. Arima, F. Iachello, I. Talmi, Phys. Lett. B 76 (1978) 139; J.N. Ginocchio, Ann. Phys. 126 (1980) 234.

23. F. Iachello, I. Talmi, Rev. Mod. Phys. 59 (1987) 339.

24. D.J. Rowe, C. Bahri, W. Wijesundera, Phys. Rev. Lett. 80 (1998) 4394.

25. N. Shimizu, T. Otsuka, T. Mizusaki, M. Honma, Phys. Rev. Lett. 86 (2001) 1171; T. Otsuka, M. Honma, T. Mizusaki, N. Shimizu, Y. Utsuno, Prog. Part. Nucl. Phys. 47 (2001) 319; N. Shimizu, T. Otsuka, T. Mizusaki, M. Honma, Phys. Rev. C 70 (2004) 054313.

26. K. Kaneko, M. Hasegawa, T. Mizusaki, Phys. Rev. C 70 (2004) 051301(R); M. Hasegawa, K. Kaneko, T. Misusaki, Y. Sun, Phys. Lett. B 656 (2007) 51.

27. J.N. Ginocchio, Phys. Rev. C 71 (2005) 064325.

28. Y.A. Luo, F. Pan, T. Wang, P.Z. Ning, J.P. Draayer, Phys. Rev. C 73 (2006) 044323; Y.A. Luo, Y. Zhang, X.F. Meng, F. Pan, J.P. Draayer, Phys. Rev. C 80 (2009) 014311.

29. N. Marginean, et al., Phys. Lett. B 633 (2006) 696.

30. Y. Sun, P.M. Walker, F.R. Xu, Y.X. Liu, Phys. Lett. B 659 (2008) 165.

31. S. Cwiok, P.-H. Heenen, W. Nazarewicz, Nature 433 (2005) 705.

32. J. Meng, W. Zhang, S.Q. Zhang, H. Toki, L.S. Geng, Eur. Phys. J. A 25 (2005) 23; T. Niki c, D. Vretenar, G.A. Lalazissis, P. Ring, Phys. Rev. Lett. 99 (2007) 092502.

33. W. M. Zhang, D. H. Feng and J. N. Ginocchio, Phys. Rev. Lett. **59**, 2032(1987).

34. W. M. Zhang, D. H. Feng and J. N. Ginocchio, Phys. Rev. **C 37**, 1281(1988).

35. D. J. Rowe, C. Bahri and W. Wijesundera, Phys. Rev. Lett. **80**, 4394(1998).

36. C. Bahri, D. J. Rowe and W. Wijesundera, Phys. Rev. **C 58**, 1539(1998).

37. Y. X. Liu, Z. F. Hou, Y. Zhang abd H. Wei, arXiv:nucl-th/0611035v1.

38. J. N. Ginocchio, Phys. Rev. **C 71**, 064325(2005).

39. Z. F. Hou, Y. Zhang and Y. X. Liu, Physics Letters **B 688** (2010) 298304.

40. D.Bonatsos, E.A.McCutchan,R. F.Casten, andR. J.Casperson, Phys. Rev. Lett. 100, 142501 (2008).

41. J. Q. Chen, Nucl. Phys. **A 626**, 686(1997).

42. Y. A. Luo, F. Pan, C. Bahri, and J. P. Draayer, Phys. Rev. **C 71**, 044304(2005).

43. L.Li, Y. A. Luo, T. Wang, F. Pan and J. P. Draayer, J. Phys. G: Nucl. Part. Phys. **36**(2009)125107.

44. Y. A. Luo, F. Pan, T. Wang, P. Z. Ning and J. P. Draayer, Phys. Rev. **C 73**, 044323(2006).

45. Y. A. Luo, Y. Zhang, X. F. Meng, F. Pan and J. P. Draayer, Phys. Rev. **C 80**,014311(2009).

46. Y. Wang, L. Li, Y.A. Luo, Y. Zhang, F. Pan and J. P. Draayer, Int. J. MOd. Phys. E**20**(2011)2229.

47. Y. A. Luo, C. Bahri, F. Pan, V. G. Gueorguiev and J. P. Draayer, Int. J. Mod. Phys. E **14**(2005)1023.

The $\eta K\bar{K}$ and $\eta' K\bar{K}$ systems with the fixed center approximation to Faddeev equations

Wei-Hong Liang[1, *], C. W. Xiao[2] and E. Oset[2]

[1] *Department of Physics, Guangxi Normal University, Guilin 541004, China*
[2] *Departamento de Física Teórica and IFIC, Centro Mixto Universidad de Valencia-CSIC, Institutos de Investigación de Paterna, Apartado 22085, 46071 Valencia, Spain*
* E-mail: liangwh@gxnu.edu.cn

We investigate the three-body systems of $\eta K\bar{K}$ and $\eta' K\bar{K}$, by taking the fixed center approximation to Faddeev equations. We find a clear and stable resonance structure around 1490 MeV in the scattering amplitude for the $\eta K\bar{K}$ system, which is not sensitive to the renormalization parameters. This resonance is associated to the $\eta(1475)$. We get only an enhancement effect of the threshold in the $\eta' K\bar{K}$ amplitude.

Keywords: Three-body system; Faddeev equation; the fixed center approximation.

1. Introduction

The three-body interaction is a subject in hadron physics drawing much attention for a long time.[1,2] Taking the Fixed Center Approximation (FCA) to the Faddeev equations provides an effective tool to deal with three-body hadron interactions.[3,4] This method is technically simple and accurate when dealing with bound states.[5,6]

In this work we will use the FCA to Faddeev equations to investigate the $\eta K\bar{K}$ and $\eta' K\bar{K}$ systems. When studied in S-wave, provided the strength of the interaction allows for it, these systems could give rise to η states. There are many η excited states in the Particle Data Group (PDG).[7] Since we do not want states too far from threshold, the $\eta(1475)$ could be in principle a candidate for the $\eta K\bar{K}$ system. For the $\eta' K\bar{K}$ system, all the η states are far away from the threshold. Recently, BESIII announced two resonances below 1900MeV in the J/ψ decay processes.[8,9] We shall explore the possible molecular structure of $\eta K\bar{K}$ and $\eta' K\bar{K}$ three-body systems.

2. Formalism

The FCA to Faddeev equations assumes a pair of particles (1 and 2) forming a cluster. Then particle 3 interacts with the components of the cluster, undergoing all possible multiple scattering with those components. The total three-body scattering amplitude T reads

$$T = T_1 + T_2, \qquad T_1 = t_1 + t_1 G_0 T_2, \qquad T_2 = t_2 + t_2 G_0 T_1, \qquad (1)$$

where the amplitudes t_1 and t_2 represent the unitary two-body scattering amplitudes with coupled channels for the interactions of particle 3 with particle 1 and 2, respectively, and G_0 is the propagator of particle 3 between the components of the two-body system. In our case we will take the $K\bar{K}$ forming a cluster of the $f_0(980)$, as appears in the chiral unitary approach,[10] and the η or η' will be the particle 3. The function G_0 is given by

$$G_0(s) = \frac{1}{2M_R} \int \frac{d^3\vec{q}}{(2\pi)^3} F_R(q) \frac{1}{q^{02} - \vec{q}\,^2 - m_3^2 + i\,\epsilon}, \qquad (2)$$

with $F_R(q)$ being the form factor of the cluster.[11]

The two-body scattering amplitude satisfies the coupled-channel Bethe-Salpeter equation,

$$t = [1 - VG]^{-1}V, \qquad (3)$$

where V is a matrix of the interaction potentials between the coupled channels πK, ηK and $\eta' K$, which can be found in Ref.,[12] and G is a diagonal matrix of the loop function of two mesons in the i-channel, given by[11]

$$G_i(s) = i \int \frac{d^4 q}{(2\pi)^4} \frac{1}{(P-q)^2 - m_1^2 + i\varepsilon} \frac{1}{q^2 - m_2^2 + i\varepsilon}. \qquad (4)$$

We deal with the divergent G_i in a dimensional regularization scheme, where the regularization parameters are the scale μ and the subtraction constant $a(\mu)$, although they are correlated and there is only one free parameter. We take $\mu = m_K$ and $a(\mu) = -1.383$ by fitting the experimental data of the $K\pi$ phase shifts.[12]

3. Results and Discussions

According to the formalism above, we can evaluate the total three-body scattering amplitude T of the $\eta K\bar{K}$ ($\eta'K\bar{K}$) system. The results are shown in Fig. 1. In Fig. 1 (left), we can see a clear resonance structure in the modulus squared of the $\eta K\bar{K}$ scattering amplitude, which is around 1490 MeV, with the width of about 100 MeV, and about 38 MeV below the threshold of $\eta f_0(980)$. This result is consistent with the one found in Ref.[13] From the PDG,[7] this resonance may be the $\eta(1475)$ of $I = 0$, with mass 1476 ± 4 MeV and width 85 ± 9 MeV.

We also see an obvious peak around 1940 MeV in Fig. 1 (right) for the $\eta'K\bar{K}$ interaction, which is very close to threshold, 1942 MeV. This peak should be an enhancement effect of the threshold, a cusp effect, and we will check it further.

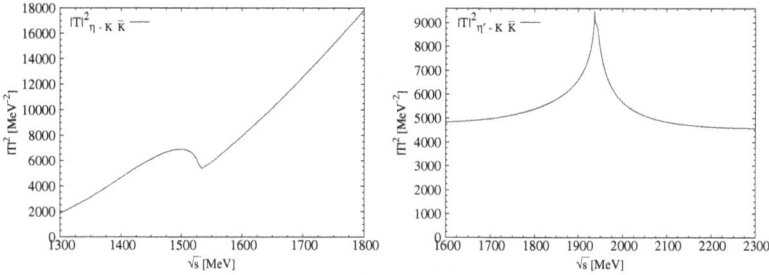

Fig. 1.   Modulus squared of the three-body interaction amplitudes. Left: $|T_{\eta K\bar{K}}|^2$; Right: $|T_{\eta'K\bar{K}}|^2$.

As a further check, we take into account the width of the $f_0(980)$ in the three-body scattering amplitudes, and find that the effects of the cluster's width are small and do not change the relevant features shown above.

Next, we check the uncertainty coming from the renormalization parameter $a(\mu)$, which is fixed by fitting the experimental data of the $K\pi$ phase shifts. Then, we change 50% up and down the $a(\mu)$, to a point where the $K\pi$ phase shifts are not too good, as shown in Fig. 2 (left). From Fig.

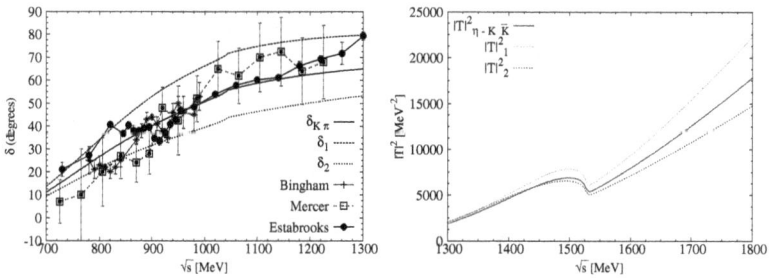

Fig. 2.   The results with the change of $a(\mu)$. Left: the $K\pi$ phase shifts, solid line is the fit one, the two dash line of δ_1, δ_2 are with 50% changes; Right: $|T_{\eta K\bar{K}}|^2$, solid line is with fit parameter, the two dash line of T_1, T_2 are with 50% changes of the fit.

2 (right), we can see that the resonance structure in $\eta K\bar{K}$ scattering is not changed so much even with these extreme changes in the input. This gives us confidence that the results that we get are rather solid. The same changes only affect in a minor way the $\eta'K\bar{K}$ amplitude and the cusp effect at threshold is the only relevant feature of the amplitude.

4. Summary

We investigate the three-body systems of $\eta K\bar{K}$ and $\eta'K\bar{K}$ by taking the fixed center approximation to Faddeev equations. We find a clear and stable resonance structure around 1490 MeV in the scattering amplitude for the $\eta K\bar{K}$ system, which is associated to the $\eta(1475)$. We get only an enhancement effect of the threshold in the $\eta'K\bar{K}$ amplitude.

Acknowledgments

We thank J. A. Oller, M. Albaladejo, F. K. Guo and A. Martinez Torres for useful discussions. This work is partly supported by the NSFC (Grant No. 11165005) and by the Spanish Ministerio de Economia y Competitividad and European FEDER funds under the Contract No. FIS2011-28853-C02-01 and the Generalitat Valenciana in the program Prometeo, 2009/090.

References

1. L. D. Faddeev, Sov. Phys. JETP **12**, 1014 (1961) [Zh. Eksp. Teor. Fiz. **39**, 1459 (1960)].
2. J. M. Richard, Phys. Rept. **212**, 1 (1992).
3. S. S. Kamalov, E. Oset and A. Ramos, Nucl. Phys. A **690**, 494 (2001).
4. A. Gal, Int. J. Mod. Phys. A **22**, 226 (2007).
5. M. Bayar, J. Yamagata-Sekihara and E. Oset, Phys. Rev. C **84**, 015209 (2011).
6. C. W. Xiao, M. Bayar and E. Oset, Phys. Rev. D **84**, 034037 (2011).
7. J. Beringer *et al.* [Particle Data Group Collaboration], Phys. Rev. D **86**, 010001 (2012).
8. M. Ablikim *et al.* [BESIII Collaboration], Phys. Rev. Lett. **107**, 182001 (2011).
9. M. Ablikim *et al.* [BESIII Collaboration], Phys. Rev. Lett. **106** (2011) 072002.
10. J. A. Oller and E. Oset, Nucl. Phys. A **620**, 438 (1997) [Erratum-ibid. A **652**, 407 (1999)].
11. Weihong Liang, C. W. Xiao and E. Oset, Phys. Rev. D **88**, 114024 (2013).
12. F. K. Guo, R. G. Ping, P. N. Shen, H. C. Chiang and B. S. Zou, Nucl. Phys. A **773**, 78 (2006).
13. M. Albaladejo, J. A. Oller and L. Roca, Phys. Rev. D **82**, 094019 (2010).

Heavy quark spin structure in hidden charm molecules

Yan-Rui Liu

School of Physics, Shandong University, Jinan, Shandong 250100, China
E-mail: yrliu@sdu.edu.cn

Recently observed heavy quark exotic XYZ mesons stimulated heated discussions about their structures. In the molecule picture, we discuss the spin of the heavy quark pair in an S-wave meson-antimeson state, whose value is either 1 or 0. We find two rules that the spin is only 1 in the heavy quark limit. From the conservation of the $c\bar{c}$ spin, one may analyze the charmonium products from strong decays. The selection rules give constraints on the products.

Keywords: Exotic state; molecule; heavy quark; selection rule.

1. Introduction

The quark model (QM)[1] is very successful in the study of hadron spectra. However, quite a few unexpected mesonic states above the $D\bar{D}$ threshold have been observed recently. They are difficult to understand in QM. Due to the closeness of the mass to some threshold, molecular interpretation is very popular. For example, $X(3872)$ is widely interpreted as a $D\bar{D}^*$ molecule while one may partly explain $Y(4260)$'s properties by assuming that it is a $D\bar{D}_1$ molecule. In the unexpected states, the observation of charged charmonium-like mesons is quite interesting.[2–5] The quark content must be 4 or more. Although these states were not predicted in QM, their existence is allowed in QCD. Up to now, molecules except the deuteron have not been confirmed. These observations give us good chances to find out hadronic molecules.

With the development of experimental measurements, more near-threshold exotic states and molecule candidates are expected. It is interesting to study the molecule picture deeper. In the charmed meson-antimeson state, the $c\bar{c}$ pair has a spin either 1 or 0. In the infinite mass limit, the $c\bar{c}$ spin is conserved in its strong decays. For a general molecule, both 1 and 0 are allowed. However, in some cases, the spin can only be 1, which is called heavy quark spin selection rule.[6] In Ref. 6, Voloshin studied the $c\bar{c}$ spin in the 1^{++} $D\bar{D}^*$ molecule and found that the spin is only 1. Here we extend that study to a general S-wave heavy quark meson-antimeson system.[7]

2. Heavy Quark Spin Structure

Heavy-light mesons are classified according to the heavy quark spin symmetry. The ground states form a $(0^-, 1^-)$ doublet with the angular momentum of light degree being $1/2$. The first orbitally excited mesons form a $(0^+, 1^+)$ doublet and a $(1^+, 2^+)$ doublet. The total angular momenta of the light degree of freedom are $1/2$ and $3/2$, respectively. The mesons in the same doublet are degenerate in the limit of infinitely heavy quark mass.

To study the $c\bar{c}$ spin in the system of a meson and an antimeson, we construct explicitly the spin wavefunction. To do that, we treat the heavy-light meson as a heavy quark with spin $j_c = 1/2$ and an "effective" antiquark with spin $\bar{j}_q$. After recouplling the meson-antimeson spin wavefunction to a $c\bar{c} - \bar{q}q$ form, one extracts the information for the $c\bar{c}$ spin. Before recoupling, we have

$$\chi_J = \frac{1}{\sqrt{2}} \left\{ (j_1 \bar{j}_2)_{J_{12}} (\bar{j}_3 j_4)_{J_{34}} + C_X (-1)^{J-J_{12}-J_{34}} (j_3 \bar{j}_4)_{J_{34}} (\bar{j}_1 j_2)_{J_{12}} \right\}. \quad (1)$$

where j $(\bar{j})$ is the spin of a quark (antiquark), J_{12} (J_{34}) is the total spin of the heavy-light meson A (B). In this $A\bar{B} \pm B\bar{A}$ wavefunction, C_X is the given C-parity of the system. In the case of $A = B$, an additional factor $\frac{1}{\sqrt{2}}$ is needed. After recoupling the wave function, one obtains

$$\chi_J = \frac{1}{\sqrt{2}} \sum_{J_{13}, J_{24}} (j_1 \bar{j}_3)_{J_{13}} [(\bar{j}_2 j_4)_{J_{24}} - C_X (-1)^{j_2 + j_4 + J_{13} + J_{24}} (\bar{j}_4 j_2)_{J_{24}}]$$

$$\times \sqrt{(2J_{12}+1)(2J_{34}+1)(2J_{13}+1)(2J_{24}+1)} \left\{ \begin{matrix} \frac{1}{2} & j_2 & J_{12} \\ \frac{1}{2} & j_4 & J_{34} \\ J_{13} & J_{24} & J \end{matrix} \right\}. \quad (2)$$

Here, $(j_3 \bar{j}_1)_{J_{13}} = (j_1 \bar{j}_3)_{J_{13}}$ is the spin wave function of the $c\bar{c}$ pair and the part in the brackets is the spin wave function of $\bar{q}\tilde{q}$ with the C-parity $c_q = C_X (-1)^{J_{13}}$. In the case of $A = B$, $(\bar{j}_2 j_4)_{J_{24}} = (\bar{j}_4 j_2)_{J_{24}}$. With the latter formula, it is not difficult to get the $c\bar{c}$ spin and corresponding amplitude.

By studying various meson-antimeson systems with total angular momentum J, one observes some results that the $c\bar{c}$ spin is only 1. We show the systems and results in Table 1. From that table, we summarize two rules for the $c\bar{c}$ spin:

(a) If $J = min(J_\ell) - 1$ or $J = max(J_\ell) + 1$, $J_{c\bar{c}} \neq 0$, where J_ℓ indicates the total angular momentum of the light degree of freedom;

Table 1. S-wave meson-antimeson states, quantum numbers, and selection rules for $J_{c\bar{c}} \neq 0$.

State	J^C	J_{24}	Selection rule for $S_{c\bar{c}} \neq 0$
$D\bar{D}^*/D_0^*\bar{D}_1'$	$1^\pm$	$0,1$	$J^C = 1^+$
$D^*\bar{D}^*/D_1'\bar{D}_1'$	$0^+,1^-,2^+$	$0,1$	$J = 2$
$D^*\bar{D}_1'$	$0^\pm,1^\pm,2^\pm$	$0,1$	$J = 2$
$D^*\bar{D}_1/D_1'\bar{D}_1$	$0^\pm,1^\pm,2^\pm$	$1,2$	$J = 0$
$D^*\bar{D}_2^*/D_1'\bar{D}_2^*$	$1^\pm,2^\pm,3^\pm$	$1,2$	$J = 3$
$D_1\bar{D}_2^*$	$1^\pm,2^\pm,3^\pm$	$0\sim3$	$J^C = 1^+,2^-,3^+$
$D_2^*\bar{D}_2^*$	$0^+,1^-,2^+,3^-,4^+$	$0\sim3$	$J = 4$

(b) If A and B are different but belong to the same doublet, $J_{c\bar{c}} \neq 0$ when $J^C = 1^+, 2^-, 3^+, \cdots$.

These rules can be used to any orbitally excited heavy-light meson case.[7] For a heavy quark baryon-antibaryon state, one also obtains similar selection rules. The exceptional case is for two singlet baryons where $J_{c\bar{c}} = J$.[7]

3. Strong Decays into Charmonia

The heavy quark spin symmetry implies that the spin-flip for heavy quarks is suppressed when the initial state decays. Therefore, we may use the obtained heavy quark spin selection rules to discuss the strong decays into charmonia.

For the $D\bar{D}^*$ system, four possible states may be formed, i.e. $I^G(J^{PC}) = 0^+(1^{++})$, $1^-(1^{++})$, $1^+(1^{+-})$, and $0^-(1^{+-})$. The $X(3872)$ state is related to the first possibility where the final charmonia should have $S_{c\bar{c}} = 1$ in the heavy quark limit. The third possibility is related with the observed charged $Z_c^+(3900)$. It is widely discussed in the molecular picture. In principle, if it is really a $D\bar{D}^*$ molecule or resonance, the strong decay products should include not only spin-triplet charmonia J/ψ and χ_{cJ}, but also spin-singlet η_c and h_c. However, BESIII does not observe significant signal of $Z_c^+(3900)$ in the invariant mass of $h_c\pi$.[8] This fact implies that $Z_c^+(3900)$ might be a general tetraquark state. Further conclusion needs more investigations.

The other interesting system is $D^*\bar{D}^*$ where the allowed J^{PC} are 0^{++}, 1^{+-}, and 2^{++}. BESIII observed two charged states around the threshold. One is in the $h_c\pi$ invariant mass $(Z_c^+(4020)^8)$, the other is in the $D^*\bar{D}^*$ invariant mass $(Z_c^+(4025)^9)$. If both of them are $D^*\bar{D}^*$ states, one may identify whether they are the same one from their strong decays. Since $1^{+-} \rightarrow \eta_c\rho$, $h_c\pi$, $J/\psi\pi$, the quantum numbers of $Z_c^+(4020)$ being 1^{+-} are favored and spin-singlet charmonium decay mode should also exist. If the

$J/\psi\pi$ channel could not be observed, this state might be a tetraquark containing dominantly $J_{c\bar{c}} = 0$. In either case, the channel $\eta_c\rho$ is expected. On the other hand, $Z_c^+(4025)$ has both spin-triplet and spin-singlet decay modes if its J^{PC} are 0^{++} or 1^{+-}. Because of the spin selection rule, its charmonium decay modes are dominantly spin-triplet if its J^{PC} are 2^{++}.

4. Summary

Many resonances near some meson-antimeson thresholds have been observed and quite a few of them have been confirmed. They are interesting in that their structures and decay properties are difficult to understand. The molecule picture can give some explanations. From the heavy quark symmetry, we study the heavy quark spin structure of an S-wave meson-antimeson molecule. We obtain two rules for the $c\bar{c}$ spin being only 1, which is convenient to analyze allowed charmonium decays. Of course, heavy quark spin symmetry is not strict in the real world and the rules are violated.

Acknowledgments

This project was supported by the National Natural Science Foundation of China (No. 11275115) and SRF for ROCS, SEM.

References

1. S. Godfrey and N. Isgur, Phys. Rev. D **32**, 189, (1985).
2. S.-K. Choi et al. (Belle Collaboration), Phys. Rev. Lett. **100**, 142001 (2008); R. Mizuk et al. (Belle Collaboration), Phys. Rev. D **78**, 072004 (2008).
3. M. Ablikim et al. (BESIII Collaboration), Phys. Rev. Lett. **110**, 252001 (2013).
4. Z.Q. Liu et al. (Belle Collaboration), Phys. Rev. Lett. **110**, 252002 (2013).
5. T. Xiao, S. Dobbs, A. Tomaradze, and K.K. Seth, arXiv: 1304.3036.
6. M.B. Voloshin, Phys. Lett. B **604**, 69 (2004).
7. Y.R. Liu, Phys. Rev. D **88**, 074008 (2013).
8. M. Ablikim et al. (BESIII Collaboration), arXiv: 1309.1896.
9. M. Ablikim et al. (BESIII Collaboration), arXiv: 1308.2760.

Search for deeply bound Kaonic nuclear states via ^{3}He(K^-, n) reaction at J-PARC

Y. Ma^a, S. Ajimura^m, G. Beer^b, H. Bhang^c, M. Bragadireanu^d, P. Buehler^e,
L. Busso^{f,g}, M. Cargnelli^e, S. Choi^c, C. Curceanu^h, S. Enomoto^i, D. Faso^{f,g},
H. Fujioka^j, Y. Fujiwara^k, T. Fukuda^l, C. Guaraldo^h, T. Hashimoto^k, R. S. Hayano^k,
T. Hiraiwa^m, M. Iio^n, M. Iliescu^h, K. Inoue^i, Y. Ishiguro^j, T. Ishikawa^k, S. Ishimoto^n,
T. Ishiwatari^e, K. Itahashi^a, M. Iwai^n, M. Iwasaki^{o,a*}, Y. Kato^a, S. Kawasaki^i,
P. Kienle^p, H. Kou^o, J. Marton^e, Y. Matsuda^q, Y. Mizoi^l, O. Morra^f, T. Nagae^{j†},
H. Noumi^m, H. Ohnishi^a, S. Okada^a, H. Outa^a, K. Piscicchia^h, M. Poli Lener^h,
A. Romero Vidal^h, Y. Sada^j, A. Sakaguchi^i, F. Sakuma^a, M. Sato^a, A. Scordo^h,
M. Sekimoto^n, H. Shi^k, D. Sirghi^{h,d}, F. Sirghi^{h,d}, K. Suzuki^e, S. Suzuki^n, T. Suzuki^k,
K. Tanida^c, H. Tatsuno^h, M. Tokuda^o, D. Tomono^a, A. Toyoda^n, K. Tsukada^r,
O. Vazquez Doce^{h,s}, E. Widmann^e, B. K. Wuenschek^e, T. Yamaga^i, T. Yamazaki^{k,a},
H. Yim^t, Q. Zhang^a, and J. Zmeskal^e

(J-PARC E15 Collaboration)

(a) *RIKEN Nishina Center, RIKEN, Wako, 351-0198, Japan*
(b) *Department of Physics and Astronomy, University of Victoria, Victoria BC V8W 3P6, Canada*
(c) *Department of Physics, Seoul National University, Seoul, 151-742, South Korea*
(d) *National Institute of Physics and Nuclear Engineering - IFIN HH, Romania*
(e) *Stefan-Meyer-Institut für subatomare Physik, A-1090 Vienna, Austria*
(f) *INFN Sezione di Torino, Torino, Italy*
(g) *Dipartimento di Fisica Generale, Universita' di Torino, Torino, Italy*
(h) *Laboratori Nazionali di Frascati dell' INFN, I-00044 Frascati, Italy*
(i) *Department of Physics, Osaka University, Osaka, 560-0043, Japan*
(j) *Department of Physics, Kyoto University, Kyoto, 606-8502, Japan*
(k) *Department of Physics, The University of Tokyo, Tokyo, 113-0033, Japan*
(l) *Laboratory of Physics, Osaka Electro-Communication University, Osaka, 572-8530, Japan*
(m) *Research Center for Nuclear Physics (RCNP), Osaka University, Osaka, 567-0047, Japan*
(n) *High Energy Accelerator Research Organization (KEK), Tsukuba, 305-0801, Japan*
(o) *Department of Physics, Tokyo Institute of Technology, Tokyo, 152-8551, Japan*
(p) *Technische Universität München, D-85748, Garching, Germany*
(q) *Graduate School of Arts and Sciences, The University of Tokyo, Tokyo, 153-8902, Japan*
(r) *Department of Physics, Tohoku University, Sendai, 980-8578, Japan*

---

*Spokesperson
†Co-Spokesperson

(s) *Excellence Cluster Universe, Technische Universität München, D-85748, Garching, Germany*
(t) *Korea Institute of Radiological and Medical Sciences (KIRAMS), Seoul, 139-706, South Korea*

As the latest effort to search for deeply-bound $\bar{K}$-nuclear states, E15 experiment has been carried out at K1.8 branch beam line (K1.8BR) at J-PARC. ^{3}He(K^-, N) reaction was employed to search for the simplest $\bar{K}$-nuclear bound state, K^-pp. In this proceeding, preliminary results of ^{3}He(K^-, n) spectra obtained in the first physics-run will be presented.

Keywords: $\bar{K}$N interaction; deeply bound K^-pp state; missing-mass spectrum.

1. Introduction

The property of $\Lambda(1405)$ has been a long standing puzzle for decades. The possible interpretation of $\Lambda(1405)$ as the bound state of $\bar{K}N$ has inspired the proposal of strong attractive potential between $\bar{K}$ and N.[2] Several experiments have been performed to verify this proposal. For example, KEK-E548 experiment and FINUDA collaboration reported possible signals support the existence of strong $\bar{K}N$ attractive force[3].[4] However, final state interactions and Σ channel background make it difficult to draw any conclusion based on only missing mass (KEK-E548) or invariant mass (FINUDA) measurement. In order to fully separate signals from backgrounds, a complete measurement for both missing mass and invariant mass is necessary.

Based on this idea, our J-PARC E15 experiment employs an exclusive measurement with the in-flight ^{3}He(K^-, n) reaction to pin down ambiguities regarding to the strength of $\bar{K}N$ interaction.[1] This experiment allows us to investigate the K^-pp bound state both in the formation via missing mass spectroscopy and its decay via invariant mass spectroscopy using the emitted neutron and the expected decay, $K^-pp \to \Lambda p \to \pi^-pp$, respectively. In addition to the ^{3}He(K^-, n), the ^{3}He(K^-, p) reaction is also measured with the K1.8BR spectrometer to investigate the isospin dependence of the kaon-nucleus interaction.

2. Experimental Setup

The experimental setup for E15 is illustrated in Fig.1. The main components of E15 setup include a liquid ^{3}He target system, beam line spectrometer, Central Detection System (CDS), Neutron Counter (NC) and Proton Counter (PC). The beam line spectrometer is used to analyze the beam particle momentum with 2.2 MeV/c resolution at 1 GeV/c beam momentum. Liquid ^{3}He target is placed in the center of CDS. CDS consists of a cylindrical drift chamber, scintillation hodoscope and a solenoid magnet and is in charge of the measurement of central tracks. A resolution of ~ 7.7 MeV/c^2 is estimated based on the reconstructed invariant mass of K_S^0. A bending magnet is located in the down stream of CDS to guide unreacted minus charged beam particles to the beam dump and reacted positive particles to PC. Neutral particles are detected with NC located 15 m down stream of target, whose time resolution is ~ 160 ps and detection efficiency of $\sim 35\%$ for 1 GeV/c neutron. For details of E15 spectrometer and technical parameters, please refer to our full paper for K1.8BR spectrometer commissioning.[5]

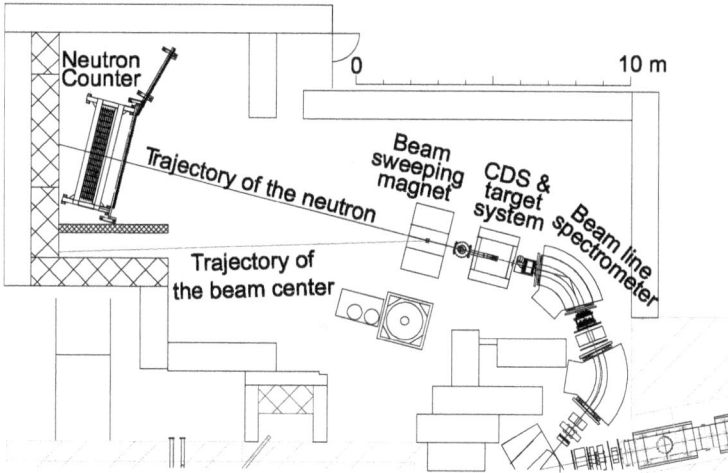

Fig. 1.   Experimental setup of J-PARC E15.

3. Preliminary Results

The first round physics run of the E15 experiment was carried out in March and May 2013. In total, 5×10^9 K^- were bombarded on the liquid ^{3}He target, which corresponds to $\sim 1\%$ of statistics requested in the original proposal.[1] Fig.2 shows the missing-mass spectrum of the ^{3}He(K, n) reaction measured by the NC. One or more charged tracks are required in the CDS to reconstruct the reaction vertex. In Fig.2, a clear peak due to the quasi-elastic $Kn \to Kn$ and the charge exchange $Kp \to K^0 n$ reactions is observed just above mass threshold of K+p+p without any potential (2.37GeV/c^2).

110

The K_S^0 tagged spectrum draw together in Fig.2 is well reproduced by a GEANT4 Monte Carlo simulation with the evaluated missing-mass resolution of $\sim$10 MeV/c². It is difficult to explain the tail structure below the K+p+p mass threshold in Fig.2 by detector effects only. Further analyses are in progress to understand the structure below the threshold.

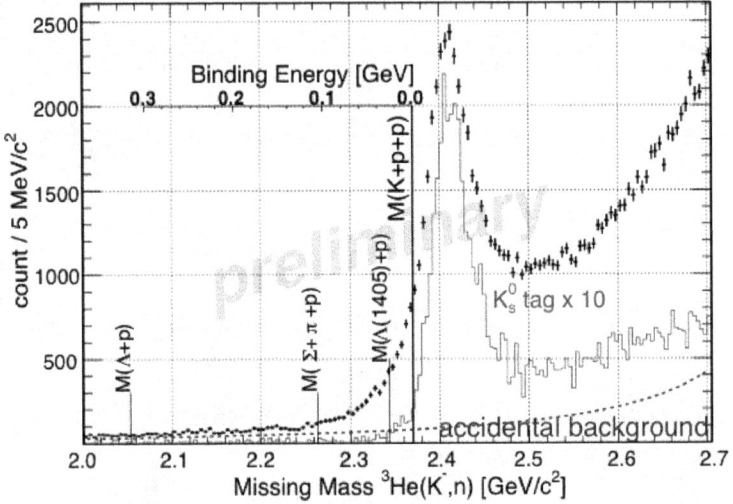

Fig. 2.   Missing mass spectrum of $^3He(K, n)$ reaction.

4. Summary

J-PARC E15 experiment is optimized for the search of K^-pp deeply bound state by performing a complete measurement for both missing mass and invariant mass with ^{3}He(K^-, n) reaction. The first physics run has been successfully carried out in 2013 and data analysis in under progress.

References

1. M. Iwasaki, et al., J-PARC E15 proposal
2. T. Yamazaki, et al., Proc. Jpn. Acad., **Ser. B 83** 144.
3. T. Kishimoto, et al., Prog. Theor. Phys., **118** 181.
4. M. Agnello, et al., Phys. Rev. Lett. **94** 212303
5. K. Agari, et al., Prog. Theor. Exp. Phys. 02B011

Studying the $e^+e^- \to (D^*\bar{D}^*)^\pm\pi^\mp$ reaction and the claim for the $Z_c(4025)$ resonance

A. Martínez Torres, K. P. Khemchandani, F. S. Navarra, M. Nielsen

Instituto de Física, Universidade de São Paulo, C.P. 66318, 05389-970 São Paulo, SP, Brazil

E. Oset

Departamento de Física Teórica and IFIC, Centro Mixto Universidad de Valencia-CSIC, Institutos de Investigación de Paterna, Aptdo. 22085, 46071 Valencia, Spain

In this talk I show the results we find for the $D^*\bar{D}^*$ invariant mass distribution associated to the reaction $e^+e^- \to (D^*\bar{D}^*)^\pm\pi^\mp$, which has been studied by the BESIII collaboration and claims of a new $J^P = 1^+$ state, called $Z_c(4025)$, have been made. As I will show, while the interpretation of the BESIII collaboration for the signal found is plausible, there are others which are equally possible, like a $J^P = 2^+$ resonance or a bound state or simply a pure D-wave background. Thus, the arguments to claim a new state around 4025 MeV in the $D^*\bar{D}^*$ invariant mass distribution get weaker.

Keywords: X; Y; Z states; hidden charm; heavy mesons spectroscopy.

1. Introduction

Recently,[1] the BESIII collaboration has studied the reaction $e^+e^- \to (D^*\bar{D}^*)^\pm\pi^\mp$ at a center of mass energy of $\sqrt{s} = 4.26$ GeV and a peak around 4026 MeV with isospin 1, and width of 20-30 MeV, called $Z_c(4025)$, has been reported from the $D^*\bar{D}^*$ invariant mass distribution. Assuming that the $D^*\bar{D}^*$ pair is produced in S-wave, the BES collaboration attributes to the state the spin-parity $J^P = 1^+$, although other assignments can not be excluded.

Several theoretical interpretations for the signal found in the $D^*\bar{D}^*$ spectrum as, for example, tetra quark states with spin-parity 1^+ or 2^+ , 1^+ $D^*\bar{D}^*$ bound states, etc., have been proposed on the basis of QCD sum rules, effective field theories implementing the heavy quark spin symmetry, and pion exchange models.[2–8] All of them find a mass compatible with the one of the $Z_c(4025)$, however, with a large uncertainty, to the point that the mass obtained is compatible with a resonance as well as with a bound state. The question which arises is if the signal found in the $e^+e^- \rightarrow (D^*\bar{D}^*)^\pm\pi^\mp$ reaction could correspond to a bound state in the $D^*\bar{D}^*$ subsystem instead of a resonance.

Indeed, in Ref. 9 a state with mass around 3920 MeV width of 120 MeV, isospin $I = 1$ and spin-parity 2^+ was predicted as a consequence of the dynamics involved in the $D^*\bar{D}^*$ system and coupled channels. This state could be very well responsible for the signal found by the BES collaboration. In this case, to get the right quantum numbers, the state should be produced in D-wave.

With the purpose of understanding the signal found by the BES collaboration, we calculate the $D^*\bar{D}^*$ invariant mass distribution associated with the $e^+e^- \rightarrow (D^*\bar{D}^*)^\pm\pi^\mp$ reaction considering different cases: 1) Production of a 1^+ state in the $D^*\bar{D}^*$ system in S-wave with the pion. 2) Production of a 2^+ state in the $D^*\bar{D}^*$ system in D-wave with the pion. 3) A pure D-wave background.

2. Formalism

In the production of a $D^*\bar{D}^*$ state we shall assume that the amplitude is a function of the $D^*\bar{D}^*$ invariant mass, $M_{D^*\bar{D}^*}$, as done in Ref. 1. In this case the differential cross section is given by[10]

$$\frac{d\sigma}{dM_{D^*\bar{D}^*}} \propto \frac{m_e^2}{s\sqrt{s}} p\tilde{q} |T|^2 F_L, \tag{1}$$

where p is the pion momentum in the e^+e^- center of mass (CM) frame and $\tilde{q}$ is the D^* momentum in the $D^*\bar{D}^*$ CM frame:

$$p = \frac{\lambda^{1/2}(s, m_\pi^2, M_{D^*\bar{D}^*}^2)}{2\sqrt{s}}, \tag{2}$$

$$\tilde{q} = \frac{\lambda^{1/2}(M_{D^*\bar{D}^*}^2, m_{D^*}^2, m_{\bar{D}^*}^2)}{2M_{D^*\bar{D}^*}}. \tag{3}$$

The factor $F_L = p^{2L}$ accounts for the partial wave in which the $D^*\bar{D}^*$ system is produced ($L = 0$ for S-waves and $L = 2$ for D-waves), and T is an amplitude which we parametrize as

$$T = \frac{A}{M_{D^*\bar{D}^*}^2 - M_R^2 + iM_R\Gamma_R}, \quad A \equiv \text{constant} \tag{4}$$

for the case of a state produced with mass M_R and width Γ_R. In case of a pure phase space, T would be taken as a constant.

In general, as done in Ref. 1, the $D^*\bar{D}^*$ invariant mass distribution can have contribution from a small background proportional to the phase space, and from combinatorial backgrounds (estimated by combining a reconstructed D^+ with a pion of the wrong charge, see Ref. 1 for more details) referred to as wrong sign (WS) background. Thus, we can write

$$\frac{d\sigma}{dM_{D^*\bar{D}^*}} = \frac{m_e^2}{s\sqrt{s}}p\tilde{q}\left(|T|^2 F_L + B\right) + \text{WS}, \tag{5}$$

where B represents a constant.

To determine Eq. (5) we have four unknown parameters: the mass of the state, M_R, its width, Γ_R, the magnitude of the resonant amplitude (A) and that of the phase space background (B). The strategy followed to constrain these parameters consists of performing a fit to the experimental data with the mentioned degrees of freedom and minimize the χ^2 per degrees of freedom to obtain a value of ~ 1, as done in Ref. 1.

3. Results

3.1. $D^*\bar{D}^*$ spectrum

In Fig. 1, left panel, we show the result for the $D^*\bar{D}^*$ spectrum for the case of a 1^+ state produced in S-wave with respect to the pion, with a mass around 4030 MeV and a width of 34 MeV, values which are similar to the ones adopted by the BES collaboration. In the right panel we show the same distribution but for the case of a bound state with mass 3990 MeV, width of 160 MeV (similar to the result found in Ref. 9), which is produced in D-wave with respect to the pion. As can be seen, both solutions can perfectly explain the data points. We have also found that a 2^+ resonance and a pure D-wave background are also compatible with the data found by the BES collaboration. For more details and discussions see Ref. 11.

3.2. Energy dependence of the $D^*\bar{D}^*$ spectrum

In Fig. 2 we show the $D^*\bar{D}^*$ invariant mass distributions of Fig. 1 for three values of $\sqrt{s}$, 4.26, 4.4 and 4.6 GeV. To compare them, we have renormalized the results associated to the energies $\sqrt{s} = 4.4$ GeV and $\sqrt{s} = 4.6$ GeV to the one of $\sqrt{s} = 4.26$ GeV. As can be seen, for the case of the 1^+ state not much changes in the $D^*\bar{D}^*$ invariant mass spectrum when varying $\sqrt{s}$, while for the case of the 2^+ state the changes in the distribution are bigger.

In summary, we have calculated the $D^*\bar{D}^*$ spectrum related to the reaction $e^+e^- \to (D^*\bar{D}^*)^\pm\pi^\mp$ studied by the BES collaboration and obtained

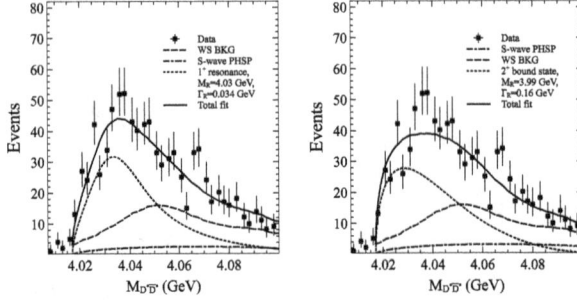

Fig. 1. (Left panel) Invariant $D^*\bar{D}^*$ mass distribution obtained using Eq. (5) for the case of a 1^+ resonance in relative S-wave with respect to pion. We use the following nomenclature in the figures: PHSP means phase space, BKG means background and WS means wrong sign. (Right panel) Same but for the case of a 2^+ $D^*\bar{D}^*$ bound state in D-wave with the pion.

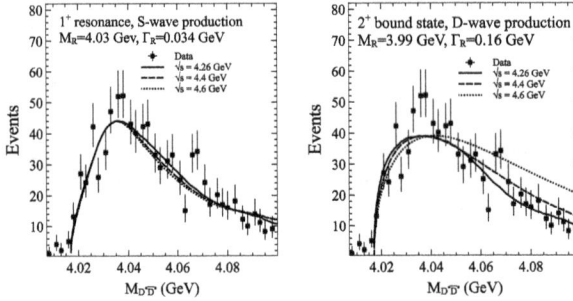

Fig. 2. Same as in Fig. 1 but for different $\sqrt{s}$ values.

that there are several possible explanations for the signal found. By studying the energy dependence of the $D^*\bar{D}^*$ spectrum obtained for each case, we arrive to the conclusion that it is possible to know if the origin of the signal found is due to a resonance or a bound state.

References

1. M. Ablikim *et al.* [BESIII Collaboration], arXiv:1308.2760 [hep-ex].
2. W. Chen, T. G. Steele, M. -L. Du and S. -L. Zhu, arXiv:1308.5060 [hep-ph].
3. C. -Y. Cui, Y. -L. Liu and M. -Q. Huang, arXiv:1308.3625 [hep-ph].
4. K. P. Khemchandani, A. Martinez Torres, M. Nielsen and F. S. Navarra, arXiv:1310.0862 [hep-ph].
5. J. He, X. Liu, Z. -F. Sun and S. -L. Zhu, arXiv:1308.2999 [hep-ph].
6. X. Wang, Y. Sun, D. -Y. Chen, X. Liu and T. Matsuki, arXiv:1308.3158 [hep-ph].

7. C. -F. Qiao and L. Tang, arXiv:1308.3439 [hep-ph].
8. C. W. Xiao, J. Nieves and E. Oset, Phys. Rev. D **88**, 056012 (2013).
9. R. Molina and E. Oset, Phys. Rev. D **80**, 114013 (2009).
10. D. Gamermann and E. Oset, Eur. Phys. J. A **36**, 189 (2008).
11. A. Martinez Torres, K. P. Khemchandani, F. S. Navarra, M. Nielsen and E. Oset, arXiv:1310.1119 [hep-ph].

Life on earth – An accident?
Chiral symmetry and the anthropic principle

Ulf-G. Meißner [for the NLEFT Collaboration]

HISKP and BCTP, Bonn University, D-53115 Bonn, Germany
IAS, IKP and JCHP, Forschungszentrum Jülich D-52425 Jülich, Germany
E-mail: www.itkp.uni-bonn.de/~meissner
E-mail: meissner@hiskp.uni-bonn.de

I discuss the fine-tuning of the nuclear forces and in the formation of nuclei in the production of the elements in the Big Bang and in stars.

1. Definition of the Problem

The elements that are pertinent to life on Earth are generated in the Big Bang and in stars through the fusion of protons, neutrons and nuclei. In Big Bang nucleosynthesis (BBN), alpha particles (^{4}He nuclei) and some heavier elements are generated. Life essential elements like ^{12}C and ^{16}O are generated in hot, old stars, where the so-called triple-alpha reaction plays an important role. Here, two alphas fuse to produce the instable, but long-lived ^{8}Be nucleus. As the density of ^{4}He nuclei in such stars is high, a third alpha fuses with this nucleus before it decays. However, to generate a sufficient amount of ^{12}C and ^{16}O, an excited state in ^{12}C at an excitation energy of 7.65 MeV with spin zero and positive parity is required as pointed out by Hoyle long ago.[1] In a further step, carbon is turned into oxygen without such a resonant condition. So we are faced with a multitude of fine-tunings which need to be explained. We know that all strongly interacting composites like hadrons and nuclei must emerge from the underlying gauge theory of the strong interactions, Quantum Chromodynamics (QCD), that is formulated in terms of quarks and gluons. These fundamental matter and force fields are, however, confined. Further, the mass of the light quarks relevant for nuclear physics is very small and thus plays little role in the total mass of nucleons and nuclei. Finally, protons and neutrons form nuclei. This requires the inclusion of electromagnetism, characterized by the fine-structure constant $\alpha_{EM} \simeq 1/137$. So the question we want to

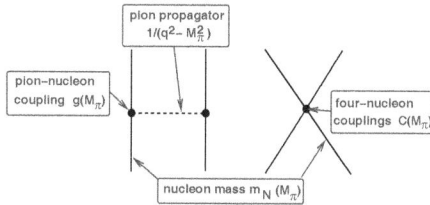

Fig. 1. Explicit and implicit pion (quark) mass dependence of the leading order nucleon–nucleon (NN) potential. Solid (dashed) lines denote nucleons (pions).

address in the following is: How sensitive are these strongly interacting composites to variations in the fundamental parameters of QCD+QED? or stated differently: how accidental is life on Earth?

2. The Nuclear Force at Varying Quark Mass

Nuclear forces are best described by utilizing chiral effective field theory (EFT) as pioneered by Weinberg.[2] The forces between two, three and four nucleons are given by pion-exchange contributions and short-distance multi-nucleon operators, the latter being accompanied by low-energy constants that must be determined by a fit to data. For a review, see Ref. 3. In this scheme, the quark mass dependence of the forces is generated explicitly (pion propagator) and implicitly (pion-nucleon coupling, nucleon mass, 4N couplings), see Fig. 1. Throughout, we use the Gell-Mann–Oakes–Renner relation, $M_\pi^2 \sim (m_u + m_d)$, so one can use pion and quark mass dependence synonymously. For any observable $\mathcal{O}$ of a hadron H, we can define its quark mass dependence in terms of the so-called K-factor, $\delta\mathcal{O}_H/\delta m_f \equiv K_H^f (\mathcal{O}_H/m_f)$, with $f = u, d, s$, and m_f the corresponding quark mass. The pion mass dependence of pion and nucleon properties can be obtained from lattice QCD combined with chiral perturbation theory as detailed in Ref. 4. The pertinent results are: $K_{M_\pi}^q = 0.494^{+0.009}_{-0.013}$, $K_{F_\pi}^q = 0.048 \pm 0.012$, and $K_{m_N}^q = 0.048^{+0.002}_{-0.006}$, where q denotes the average light quark mass. For the quark mass dependence of the short-distance terms, one has to resort to modeling using resonance saturation.[5] This induces a sizeable uncertainty that might be overcome by lattice simulations in the future. For the NN scattering lengths, this leads to $K_{1S0}^q = 2.3^{+1.9}_{-1.8}$, $K_{3S1}^q = 0.32^{+0.17}_{-0.18}$ and $K_{BE(deut)}^q = -0.86^{+0.45}_{-0.50}$ (with BE denoting the binding energy), extending and improving earlier work based on EFTs and models.[6–10] The running of the NN scattering lengths and the deuteron BE with the light quark mass is shown in Fig. 2. In addition to shifts in m_q,

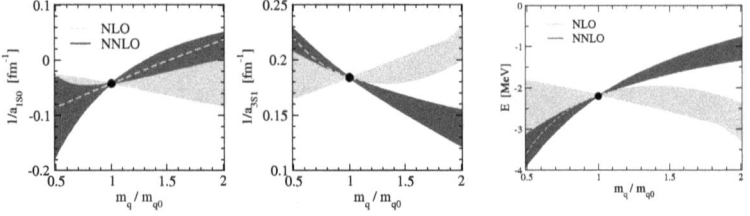

Fig. 2. Quark mass dependence of the inverse scattering length $1/a_{1S0}$ and $1/a_{3S1}$ and the deuteron binding energy. Here, m_{q0} denotes the physical light quark mass.

we shall also consider the effects of shifts in α_{EM}. The treatment of the Coulomb interaction in the nuclear lattice EFT framework is described in detail in Ref. 11.

3. Constraints from Big Bang Nucleosynthesis

With the results from the previous section, one can now analyze the constraints of element abundances in BBN on possible quark mass variations. To answer this question, we also need the variation of the ^{3}He and ^{4}He BEs with the pion mass. Following Ref. 12 (BLP), these can be obtained by convoluting the 2N K-factors with the variation of the 3- and 4-particle BEs with respect to the singlet and triplet NN scattering lengths. This gives $K^q_{3He} = -0.94 \pm 0.75$ and $K^q_{3He} = -0.55 \pm 0.42$,[4] which is consistent with a direct calculation using nuclear lattice simulations, $K^q_{3He} = -0.19 \pm 0.25$ and $K^q_{3He} = -0.16 \pm 0.26$.[13] With this input, we can calculate the BBN response matrix of the primordial abundances Y_a at fixed baryon-to-photon ratio, $\delta Y_a / \delta m_q = \sum_{X_i} (\delta \ln Y_a / \delta \ln X_i) K^q_{X_i}$, with X_i the relevant BEs for ^{2}H, ^{3}H, ^{3}He, ^{4}He, ^{6}Li, ^{7}Li and ^{7}BE and the singlet NN scattering length, using the updated Kawano code (for details, see Ref. 14). Combining the calculated with the observed abundances, one finds that the most stringent limits arise from the deuteron abundance [deut/H] and the ^{4}He abundance normalized to the one of protons, ^{4}He(Y_p), as most neutrons end up in the alpha nucleus. Combining these leads to the constraint $\delta m_q / m_q = (2 \pm 4)\%$. In contrast to most earlier determinations, we provide reliable error estimates due to the underlying EFT. However, as pointed out by BLP, one can obtain an even stronger bound due to the neutron lifetime, which strongly affects ^{4}He(Y_p). We have re-evaluated this constraint under the model-independent assumption that *all* quark and lepton masses vary with the Higgs vacuum expectation value v, leading to

$$|\delta v/v| = |\delta m_q/m_q| \leq 0.9\% .\qquad (1)$$

4. The Fate of Carbon-based Life as a Function of the Fundamental Parameters of QCD+QED

I now turn to the central topic of this talk, namely how fine-tuned is the production of carbon and oxygen with respect to changes in the fundamental parameters of QCD+QED? Or, stated differently, how much can we detune these parameters from their physical values to still have an habitable Earth as shown in Fig. 3. To be more precise, we must specify which parameters we can vary. In QCD, the strong coupling constant is tied to the nucleon mass through dimensional transmutation. However, the light quark mass (here, only the strong isospin limit is relevant) is an external parameter. Naively, one could argue that due to the small contribution of the quark masses to the proton and the neutron mass, one could allow for sizeable variations. However, the relevant scale to be compared to here is the average binding energy per nucleon, $E/A \leq 8\,\mathrm{MeV}$ (which is much smaller than the nucleon mass). As noted before, the Coulomb repulsion

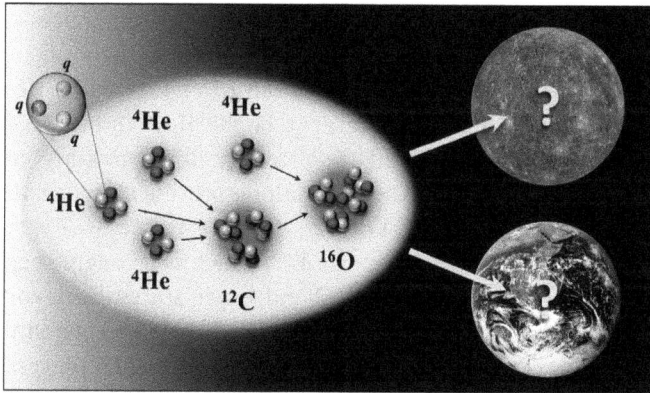

Fig. 3. Graphical representation of the question of how fine-tuned is life on Earth under variations of the average light quark mass and α_{EM}. Figure courtesy of Dean Lee.

between protons is an important ingredient in nuclear binding, therefore we must also consider changes in α_{EM}. The tool to do this are nuclear lattice simulations, which allowed e.g. for the first *ab initio* calculation of the Hoyle state.[15] Let us consider first QCD (for details, see Refs. 16 and 17). We want to calculate the variations of the pertinent energy differences in the triple-alpha process $\delta\Delta E/\delta M_\pi$, which according to Fig. 1 boils down to (we consider small variations around the physical value of the pion mass

M_π^{ph}):

$$\left.\frac{\partial E_i}{\partial M_\pi}\right|_{M_\pi^{\mathrm{ph}}} = \left.\frac{\partial E_i}{\partial \tilde{M}_\pi}\right|_{M_\pi^{\mathrm{ph}}} + x_1 \left.\frac{\partial E_i}{\partial m_N}\right|_{m_N^{\mathrm{ph}}} + x_2 \left.\frac{\partial E_i}{\partial \tilde{g}_{\pi N}}\right|_{\tilde{g}_{\pi N}^{\mathrm{ph}}}$$

$$+ x_3 \left.\frac{\partial E_i}{\partial C_0}\right|_{C_0^{\mathrm{ph}}} + x_4 \left.\frac{\partial E_i}{\partial C_I}\right|_{C_I^{\mathrm{ph}}}, \tag{2}$$

with the definitions

$$x_1 \equiv \left.\frac{\partial m_N}{\partial M_\pi}\right|_{M_\pi^{\mathrm{ph}}}, \quad x_2 \equiv \left.\frac{\partial g_{\pi N}}{\partial M_\pi}\right|_{M_\pi^{\mathrm{ph}}}, \quad x_3 \equiv \left.\frac{\partial C_0}{\partial M_\pi}\right|_{M_\pi^{\mathrm{ph}}}, \quad x_4 \equiv \left.\frac{\partial C_I}{\partial M_\pi}\right|_{M_\pi^{\mathrm{ph}}}, \tag{3}$$

with $\tilde{M}_\pi$ the pion mass appearing in the pion-exchange potentials. The various derivatives in Eq. (2) can be obtained precisely using Auxiliary Field Quantum Monte Carlo techniques and the x_i ($i = 1, 2, 3, 4$) are related to the pion and nucleon K-factors determined in Sec. 2. The scheme-dependent quantities $x_{3,4}$ can be traded for the pion-mass dependence of the inverse singlet and triplet scattering lengths, $\bar{A}_s \equiv \partial a_s^{-1}/\partial M_\pi|_{M_\pi^{\mathrm{ph}}}$, $\bar{A}_t \equiv \partial a_t^{-1}/\partial M_\pi|_{M_\pi^{\mathrm{ph}}}$. We can then express all energy differences appearing in the triple-alpha process ($\Delta E_b \equiv E_8 - 2E_4, \Delta E_h \equiv E_{12}^\star - E_8 - E_4, \varepsilon \equiv E_{12}^\star - 3E_4$, with E_4 and E_8 the energies of the ground states of ^{4}He and ^{8}Be, respectively, and $E_{12}^\star$ denotes the energy of the Hoyle state) as functions of $\bar{A}_s$ and $\bar{A}_t$. One finds that all these energy differences are correlated, i.e. the various fine-tunings in the triple-alpha process are not independent of each others, see the left panel of Fig. 4. Further, one finds a strong dependence on the variations of the ^{4}He BE, which is strongly suggestive of the α-cluster structure of the ^{8}Be, ^{12}C and Hoyle states. Such correlations related to the production of carbon have indeed been speculated upon earlier.[18,19] Consider now the reaction rate of the triple-alpha process as given by $r_{3\alpha} \sim N_\alpha^3 \Gamma_\gamma \exp(-\varepsilon/k_B T)$, with N_α the α-particle number density in the stellar plasma with temperature T, $\Gamma_\gamma = 3.7(5)$ meV the radiative width of the Hoyle state and k_B is Boltzmann's constant. The stellar modeling calculations of Refs. 20, 21 suggest that sufficient abundances of both carbon and oxygen can be maintained within an envelope of ± 100 keV around the empirical value of $\varepsilon = 379.47(18)$ keV. This condition can be turned into a constraint on shifts in m_q that reads (for more details, see Ref. 17)

$$\left|\left[0.572(19)\,\bar{A}_s + 0.933(15)\,\bar{A}_t - 0.064(6)\right] \left(\frac{\delta m_q}{m_q}\right)\right| < 0.15\% . \tag{4}$$

The resulting constraints on the values of $\bar{A}_s$ and $\bar{A}_t$ compatible with the condition $|\delta\varepsilon| < 100$ keV are visualized in the right panel of Fig. 4. The

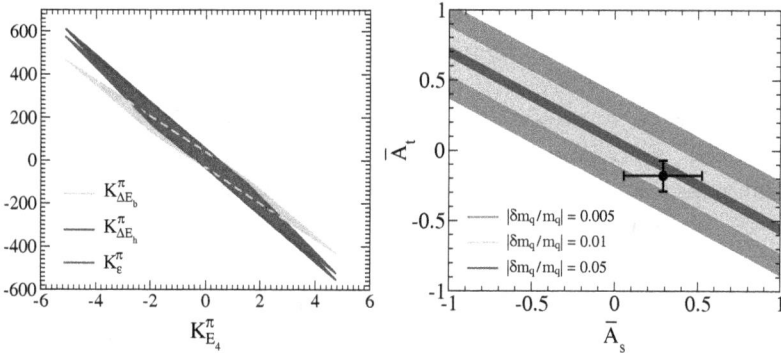

Fig. 4. Left panel: Sensitivities of ΔE_h, ΔE_b and ε to changes in M_π, as a function of $K^\pi_{E_4}$ under independent variation of $\bar{A}_s$ and $\bar{A}_t$ over the range $\{-1\ldots 1\}$. The bands correspond to ΔE_b, ε and ΔE_h in clockwise order. Right panel: "Survivability bands" for carbon-oxygen based life from Eq. (4), due to 0.5% (broad outer band), 1% (medium band) and 5% (narrow inner band) changes in m_q in terms of the input parameters $\bar{A}_s$ and $\bar{A}_t$. The most up-to-date N^{2}LO analysis of $\bar{A}_s$ and $\bar{A}_t$ from Ref. 14 is given by the data point with horizontal and vertical error bars.

various shaded bands in this figure cover the values of $\bar{A}_s$ and $\bar{A}_t$ consistent with carbon-oxygen based life, when m_q is varied by 0.5%, 1% and 5%. Given the current theoretical uncertainty in $\bar{A}_s$ and $\bar{A}_t$, our results remain compatible with a vanishing $\partial \varepsilon / \partial M_\pi$, in other words with a complete lack of fine-tuning. Interestingly, Fig. 4 (right panel) also indicates that the triple-alpha process is unlikely to be fine-tuned to a higher degree than $\simeq 0.8\%$ under variation of m_q. The central values of $\bar{A}_s$ and $\bar{A}_t$ from Ref. 14 suggest that variations in the light quark masses of up to $2-3\%$ are unlikely to be catastrophic to the formation of life-essential carbon and oxygen. A similar calculation of the tolerance for shifts in the fine-structure constant $\alpha_{\rm EM}$ suggests that carbon-oxygen based life can withstand shifts of $\simeq 2.5\%$ in $\alpha_{\rm EM}$.

5. A Short Discussion of the Anthropic Principle

The Hoyle state dramatically increases the reaction rate of the triple-alpha process. The resulting enhancement is also sensitive to the exact value of ε, which is therefore the principal control parameter of this reaction. As the Hoyle state is crucial to the formation of elements essential to life as we know it, this state has been nicknamed the "level of life".[22] Thus, the Hoyle state is often viewed as a prime manifestation of the anthropic principle, which states that the observable values of the fundamental physical and cosmological parameters are restricted by the requirement that life can form to determine them, and that the Universe be old enough for that to occur.[23,24] See, however, Ref. 25 for a thorough historical discussion of the Hoyle state in view of the anthropic principle. We remark that in the context of cosmology and string theory, the anthropic principle and its consequences have had a significant influence, as reviewed recently in Ref. 26. As noted already in Ref. 19, the allowed variations in ε are not that small, as $|\delta\varepsilon/\varepsilon| \simeq 25\%$ still allows for carbon-oxygen based life. So one might argue that the anthropic principle is indeed *not* needed to explain the fine-tunings in the triple-alpha process. However, as we just showed, this translates into allowed quark mass variations of $2 - 3\%$ and modifications of the fine-structure constant of about 2.5%. The fine-tuning in the fundamental parameters is thus much more severe than the one in the energy difference ε. Therefore, beyond such relatively small changes in the fundamental parameters, the anthropic principle indeed appears necessary to explain the observed abundances of ^{12}C and ^{16}O.

6. Summary and Outlook

In this talk, I have summarized recent developments in our understanding of the fine-tuning in the generation of the life-essential elements as well as the light elements generated in BBN. As shown, the allowed parameter variations in QCD+QED are small, giving some credit to the anthropic principle. To sharpen these conclusions, future work is required. On one side, lattice QCD at sufficiently small quark masses will eventually be able to give tighter constraints on the parameters $\bar{A}_{s,t}$ and on the other side, nuclear lattice simulations have to be made more precise to reduce the theoretical error in the binding and excitation energies and to provide *ab initio* calculations of nuclear reactions, for first steps, see Refs. 27 and 28 .

Acknowledgments

I would like to thank my NLEFT collaborators Evgeny Epelbaum, Hermann Krebs, Timo Lähde and Dean Lee for a most enjoyable collaboration. I also thank the organizers for their perfect job. Work supported in part by DFG and NSFC (Sino-German CRC 110), Helmholtz Association (contract VH-VI-417), BMBF (grant 05P12PDFTE), and the EU (HadronPhysics3 project) Computational resources provided by the Jülich Supercomputing Centre (JSC) at the Forschungszentrum Jülich and by RWTH Aachen.

References

1. F. Hoyle, Astrophys. J. Suppl. Ser. **1**, 121 (1954).
2. S. Weinberg, Phys. Lett. B **251**, 288 (1990).
3. E. Epelbaum, H.-W. Hammer, and U.-G. Meißner, Rev. Mod. Phys. **81**, 1773 (2009).
4. J. C. Berengut, E. Epelbaum, V. V. Flambaum, C. Hanhart, U.-G. Meißner, J. Nebreda, and J. R. Peláez, Phys. Rev. D **87**, 085018 (2013).
5. E. Epelbaum, U.-G. Meißner, W. Gloeckle and C. Elster, Phys. Rev. C **65**, 044001 (2002)
6. H. Müther, C. A. Engelbrecht and G. E. Brown, Nucl. Phys. A **462**, 701 (1987).
7. S. R. Beane and M. J. Savage, Nucl. Phys. A **713**, 148 (2003).
8. E. Epelbaum, U.-G. Meißner and W. Gloeckle, Nucl. Phys. A **714**, 535 (2003).
9. V. V. Flambaum and R. B. Wiringa, Phys. Rev. C **76**, 054002 (2007).
10. J. Soto and J. Tarrus, Phys. Rev. C **85**, 044001 (2012).
11. E. Epelbaum, H. Krebs, D. Lee, and U.-G. Meißner, Eur. Phys. J. A **45**, 335 (2010).
12. P. F. Bedaque, T. Luu, and L. Platter, Phys. Rev. C **83**, 045803 (2011).
13. T. Lähde, *private communication*.
14. J. C. Berengut, V. V. Flambaum and V. F. Dmitriev, Phys. Lett. B **683**, 114 (2010).
15. E. Epelbaum, H. Krebs, D. Lee, and U.-G. Meißner, Phys. Rev. Lett. **106**, 192501 (2011).
16. E. Epelbaum, H. Krebs, T. A. Lähde, D. Lee and U.-G. Meißner, Phys. Rev. Lett. **110**, 112502 (2013).
17. E. Epelbaum, H. Krebs, T. A. Lähde, D. Lee and U.-G. Meißner, Eur. Phys. J. A **49**, 82 (2013).
18. M. Livio, D. Hollowell, A. Weiss, and J. W. Truran, Nature **340**, 281 (1989).
19. S. Weinberg, "Facing Up" (Harvard University Press, Cambridge, Massachusetts, 2001).
20. H. Schlattl, A. Heger, H. Oberhummer, T. Rauscher, and A. Csótó, Astrophys. Space Sci. **291**, 27 (2004).
21. H. Oberhummer, A. Csótó, and H. Schlattl, Nucl. Phys. A **689**, 269 (2001).
22. A. Linde, "The inflationary multiverse," in *Universe or multiverse?*, edited by B. Carr (Cambridge University Press, Cambridge, England, 2007).
23. B. Carter, "Large number coincidences and the anthropic principle", in *Confrontation of cosmological theories with observational data*, edited by M. S. Longair (Reidel, Dordrecht, 1974).
24. B. J. Carr and M. Rees, Nature **278**, 605 (1979).
25. H. Kragh, Arch. Hist. Exact Sci. **64**, 721 (2010).
26. A. N. Schellekens, arXiv:1306.5083 [hep-ph].
27. G. Rupak and D. Lee, Phys. Rev. Lett. **111**, 032502 (2013).
28. M. Pine, D. Lee and G. Rupak, Eur. Phys. J. A **49**, 151 (2013).

Decays of doubly charmed meson molecules

R. Molina[1], A. Hosaka[2] and H. Nagahiro[3]

[1,2] *Research Center for Nuclear Physics (RCNP), Osaka University,*
Ibaraki, Osaka 567-0047, Japan
[3] *Department of Physics, Nara Women's University, Nara 630-8506, Japan*
[1] *E-mail: molina@rcnp.osaka-u.ac.jp*
[2] *E-mail: hosaka@rcnp.osaka-u.ac.jp*
[3] *E-mail: nagahiro@cc.nara-wu.ac.jp*

The interaction between pseudoscalar and/or vector mesons can be studied using hidden gauge Lagrangians. In this framework, the interaction between charmed mesons has been studied. Furthermore, doubly charmed states are also predicted. These new states are near the D^*D^* and $D^*D_s^*$ thresholds, and have spin-parity $J^P = 1^+$. We evaluate the decay widths of these states, named as $R_{cc}(3970)$ and $S_{cc}(4100)$ (with strangeness), and obtain 44 MeV for the non-strangeness, and 24 MeV for the doubly charm-strange state. Essentially, the decay modes are $DD_{(s)}\pi$ and $DD_{(s)}\gamma$, being the $D\pi$ and $D\gamma$ emitted by one of the D^* meson which forms the molecule.

Keywords: Doubly charm mesons; XYZ; decays of mesons; exotic mesons.

1. Introduction

Recently, the LHCb has measured the quantum numbers of the X(3872) as 1^{++}.[1] This result rules out the X(3872) to be a charmonium state, favoring the molecular interpretation.[2] In addition, several authors have discussed whether some of the other observed XYZ particles can be described in terms of molecules.[3–5] Some of the reasons on why these states cannot be accomodated into $c\bar{c}$ are the unusually high decay rates into $(\rho, \omega \text{ or} \phi)J/\psi$.[2] Also, charged states Z_c and decays between them are observed.[6]

Using hidden gauge Lagrangians combined with unitarity in coupled channels, some of the observed states which are near the open charm thresholds, are well described in terms of two-meson molecules.[3,4] Moreover, two-meson bound states of D^*D^* or $D^*D_s^*$ are dynamically generated.[7] Those doubly charmed mesons form a charged isospin singlet and doublet, they are called $R_{cc}^+(3970)$ and $S_{cc}^{+(+)}(4100)$, for the non-strangeness and strangeness one respectively. In this talk, in order to explore further the internal structure of these states, we study the decays of these states in detail.

2. Decay Modes of Doubly Charm States

The two D^* mesons can form a molecular state of spin and parity $J^P = 1^+$ when they are dominated by an s-wave state. Due to these quantum numbers, it cannot decay into $D\bar{D}$. Strong and radiative decays of the doubly charm states occur through $DD^*_{(s)}$ (or $D_{(s)}D^*$) which subsequently go to three body states via $D^* \to \pi D$ or $D\gamma$. Direct decays into three-body states, $DD\gamma$, are also evaluated, but they are small as compared to the above processes going through two bodies. The set of Feynman diagrams considered are depicted in Fig. 2. The $R_{cc} \to DD^*_{(s)}$ transition can be

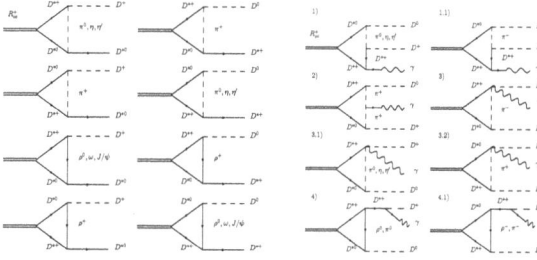

Fig. 1. Left: Feynman diagrams evaluated in the decay $R_{cc} \to DD^*$. Right: Diagrams for the $R_{cc}^+ \to D^0 D^+ \gamma$ decay through one loop.

reached through anomalous couplings VVP with pseudoscalar or vector meson exchange. The Lagrangians needed to evaluate the decay width to $DD^*_{(s)}$ are,[8]

$$\mathcal{L}_{PPV} = -ig\langle V^\mu[P, \partial_\mu P]\rangle, \qquad \mathcal{L}_{3V} = ig\langle(V^\mu\partial_\nu V_\mu - \partial_\nu V_\mu V^\mu)V^\nu\rangle$$
$$\mathcal{L}_{VVP} = \frac{G'}{\sqrt{2}}\epsilon^{\mu\nu\alpha\beta}\langle\partial_\mu V_\nu\partial_\alpha V_\alpha P\rangle \qquad (1)$$

with e the unit electronic charge, $G' = 3g'^2/(4\pi^2 f)$, $g' = -G_V M_\rho/(\sqrt{2}f^2)$, $G_V = f/\sqrt{2}$ and $g = M_V/2f$. The constant f is the pion decay constant $f = 93\ MeV$, $Q = diag(2, -1, -1, 1)/3$ and M_V is the mass of the vector meson. The P and V matrix contain the 15-plet of the pseudoscalars and vectors respectively in the physical basis. In[7] the uncertainties related with the SU(4) breaking of the coupling g are studied, considering both heavy and light couplings. These decays come through one loop which involves an integral which is logarithmically divergent, however this divergence is related to the vertex that couples the resonance to the two-meson molecular states, and is also present in the two-meson loop function, G, when those states are dynamically generated.[7] Thus, the same value of the cutoff needed to obtain these states at their masses[7] is used to evaluate the integral involved in the decays in the one-loop diagrams of Fig. 2. Once set the cutoff, one has a fixed mass and coupling of the bound state to the two-meson component, g_R. Since these three magnitudes are related, there is only one free parameter in the calculation, the cutoff q_{max}, and performing variations of this parameter one has an idea of the uncertainties in the decay widths. This is reflected in the errors of the widths, where 15% variations around its central value, 750 MeV, have been considered.

Table 1. Total and partial decay widths of the different decay modes of the doubly charmed states.

State	Channel k	Γ^k [MeV]	Channel j	Γ_j^k [MeV]	Γ_{tot} [MeV]
$R_{cc}^+(3970)$			*Hadronic decays*		
	$D^0 D^{*+}$	22 ± 6	$D^0(D^+\pi^0)$	7 ± 2	44 ± 12
			$D^0(D^0\pi^+)$	15 ± 4	
	$D^+ D^{*0}$	22 ± 6	$D^+(D^0\pi^0)$	14 ± 4	
			Radiative decays		
	$D^+ D^{*0}$		$D^+(D^0\gamma)$	8 ± 2	
	$D^0 D^{*+}$		$D^0(D^+\gamma)$	0.4 ± 0.2	
			$D^0 D^+\gamma$	$(2 \pm 1) \times 10^{-3}$	
			$D^{*0}D^+\gamma$	$(0.03 \pm 0.01) \times 10^{-3}$	
			$D^{*+}D^0\gamma$	$(0.5 \pm 0.2) \times 10^{-3}$	
$S_{cc}^+(4100)$			*Hadronic decays*		
	$D_s^+ D^{*0}$	12 ± 4	$D_s^+(D^0\pi^0)$	7 ± 2	24 ± 8
	$D^0 D_s^{*+}$	12 ± 4	-	-	
			Radiative decays		
	$D^0 D_s^{*+}$		$D^0(D_s^+\gamma)$	11 ± 4	
	$D_s^+ D^{*0}$		$D_s^+(D^0\gamma)$	5 ± 2	
			$D^0 D_s^+\gamma$	$(2 \pm 1) \times 10^{-3}$	
			$D^{*0}D_s^+\gamma$	$(0.3 \pm 0.1) \times 10^{-3}$	
			$D_s^{*+}D^0\gamma$	$(4 \pm 1) \times 10^{-3}$	
$S_{cc}^{++}(4100)$			*Hadronic decays*		
	$D_s^+ D^{*+}$	12 ± 4	$D_s^+(D^+\pi^0)$	4 ± 1	24 ± 8
			$D_s^+(D^0\pi^+)$	8 ± 3	
	$D^+ D_s^{*+}$	12 ± 4	-	-	
			Radiative decays		
	$D^+ D_s^{*+}$		$D^+(D_s^+\gamma)$	11 ± 4	
	$D_s^+ D^{*+}$		$D_s^+(D^+\gamma)$	0.2 ± 0.1	
			$D^+ D_s^+\gamma$	$(1.3 \pm 0.1) \times 10^{-4}$	
			$D^{*+}D_s^+\gamma$	$(0.3 \pm 0.1) \times 10^{-3}$	
			$D_s^{*+}D^+\gamma$	$(0.3 \pm 0.1) \times 10^{-3}$	

The diagrams included in the evaluation of the radiative decay of doubly charmed meson molecules, $R_{cc} \to DD\gamma$ are depicted in Fig. 2 (right panel), where only non-vanishing diagrams are shown.

3. Results

The results are shown in Table 1. We observe that the total widths of the doubly charmed states are (44 ± 12), (24 ± 8), and (24 ± 8) MeV for the R_{cc}^+, S_{cc}^+ and S_{cc}^{++} respectively, giving both channels (ex. $D^0 D^{*+}$ and $D^+ D^{*0}$ for the R_{cc}^+) the same contribution to the width. The direct diagrams with three/four propagators of Fig. 2, type 1), 2) and 3), lead to a very small width of the order of few KeV in the case of the $R_{cc}^+(3970)$ and $S_{cc}^+(4100)$ and 0.13 KeV for the doubly charge state, $S_{cc}^{++}(4100)$.

4. Conclusions

We have considered the possible decay modes of the doubly charmed molecules, $R_{cc}(3970)$ and $S_{cc}(4100)$, and evaluated partial decay widths to $DD_{(s)}\pi$ and $DD_{(s)}\gamma$. We find that the main source of these decays come from the decay of a $D_{(s)}^*$ meson into $D_{(s)}\pi$ or $D_{(s)}\gamma$. These decays are mediated by the exchange of one meson, vector or pseudoscalar, between the $D^* D_{(s)}^*$ pair of the molecule. The largest width comes from ρ, π and ω exchange (decreasing order) for the $R_{cc}(3970)$. Since they are not $q\bar{q}$, having a pair of cc and doubly charged, these mesons are under challenge for experiments. Hopefully, they could be observed by the LHCb or Belle.

References

1. R. Aaij et al. (LHCb Collaboration) , Phys. Rev. Lett. **110**, 222001 (2013).
2. S. Godfrey and S. L. Olsen, Ann. Rev. Nucl. Part. Sci. **58**, 51 (2008)
3. D. Gamermann and E. Oset, Phys. Rev. D **80**, 014003 (2009).
4. R. Molina and E. Oset, Phys. Rev. D **80**, 114013 (2009).
5. T. Branz, T. Gutsche and V. E. Lyubovitskij, Phys. Rev. D **80**, 054019 (2009). J. Nieves and M. P. Valderrama, Phys. Rev. D **86**, 056004 (2012). X. Liu, Z. G. Luo, Y. R. Liu and S. L. Zhu, Eur. Phys. J. C **61**, 411 (2009). G. -J. Ding, W. Huang, J. -F. Liu and M. -L. Yan, Phys. Rev. D **79**, 034026 (2009). P. G. Ortega, J. Segovia, D. R. Entem and F. Fernandez, Phys. Rev. D **81**, 054023 (2010).
6. Z. Q. Liu et al. [Belle Collaboration], Phys. Rev. Lett. **110**, 252002 (2013)
7. R. Molina, T. Branz and E. Oset, Phys. Rev. D **82**, 014010 (2010)
8. H. Nagahiro, L. Roca, A. Hosaka and E. Oset, Phys. Rev. D **79** (2009) 014015.

Chiral symmetry breaking and restoration with mixing between quarkonium and tetraquark

Tamal K. Mukherjee* and Mei Huang**

Institute of High Energy Physics, Chinese Academy of Sciences,
Beijing, 100049, China
and
Theoretical Physics Center for Science Facilities, Chinese Academy of Sciences,
Beijing, 100049, China
* E-mail: mukherjee@ihep.ac.cn
** E-mail: huangm@ihep.ac.cn

In the framework of two flavor quark-meson model we study the effect of mixing between effective quarkonium and tetraquark fields on chiral phase transition. Depending on the values of different parameters we explore different scenarios. The common feature among all of them is that the transition for quarkonium and tetraquark happening at the same temperature for all densities. We find a sufficiently strong negative cubic self interaction coupling constant of the tetraquark field can drive the chiral phase transition to first order even at zero quark chemical potential.

Keywords: Quarkonium; tetraquark; chiral phase transition.

1. Introduction

The role of chiral condensate as an order parameter for chiral phase transition is well established. But, the role of tetraquark condensate is not understood and work in this direction has recently been started.[1,2] In Ref. 1 the effect of mixing between quarkonium and tetraquark fields on the nature of $f_0(600)$ as well as the chiral phase transition is studied. On the other hand, in Ref. 2, an interesting alternate breaking of chiral symmetry in dense matter was proposed. Motivated from these studies, in this work we are going to consider the role of quarkonium, tetraquark and their mixing in chiral phase transition. The detailed version of this work can be found in Ref. 3.

2. Model

We investigate the effect of quarkonium and tetraquark mixing on chiral phase transition in the framework of quark-meson model. The Lagrangian of our model consists of fermionic field as well as effective fields having quantum numbers of scalar mesons. There are two effective fields in our model: one is for quarkonium (Φ) and the other one represents tetraquark (Φ'). The effective quarkonium and tetraquark fields interacts with the quark field via Yukawa coupling and they also interact with each other via a term $\sim Tr(\Phi')Tr(\Phi)Tr(\Phi)$. Due to space constraint we are not going to details of our model and readers are refferred to Ref. 3

Our starting point is the mean field level thermodynamical potential per unit volume at temperature T and chemical potential μ written in terms of σ (vacuum expectation value of the bare quarkonium field) and χ (vacuum expectation value of the bare tetraquark field):

$$\Omega = U(\sigma, \chi) - 2T N_c N_f \int \frac{d^3 q}{(2\pi)^3}$$
$$[\ln(1 + e^{-(E_q - \mu)/T}) + \ln(1 + e^{-(E_q + \mu)/T})], \tag{1}$$

where,

$$U(\sigma, \chi) = -\frac{1}{2} m_\Phi{}^2 \sigma^2 - \frac{1}{2} m_{\Phi'}{}^2 \chi^2 + \frac{1}{16} \lambda_1 \sigma^4 + \frac{1}{16} \lambda_2 \chi^4$$
$$+ \frac{1}{4} g_2 \chi^3 - g_1 \sigma^2 \chi + \frac{1}{2} k \sigma^2 - 2h\sigma. \tag{2}$$

In deriving the above expression for the thermodynamic potential we have neglected the ultraviolet divergent vacuum energy term[4,5]. The single particle energy is given by $E_q = \sqrt{p^2 + m_q{}^2}$ and the constituent quark mass (m_q) is given by $m_q = g_3 \sigma + g_4 \chi$. The number of colours N_c and flavors N_f of quark used in this paper are 3 and 2 respectively.

From the extremum condition of the thermodynamic potential we obtain equation of motions for σ and χ:

$$\frac{\partial \Omega}{\partial \sigma} = 0, \quad \frac{\partial \Omega}{\partial \chi} = 0. \tag{3}$$

We solve this set of coupled equation of motions Eq.(3) self consistently at each values of T and μ to determine the behavior of σ and χ and analyze the nature of chiral phase transition.

3. Results

The mixing between bare quarkonium and tetraquark fields in our model give rise to physical mesons. The vacuum properties of those mesons are used to fix the parameters in our model. For space constraints readers are requested to consult Ref.[3] for details of the parameter fixing and the

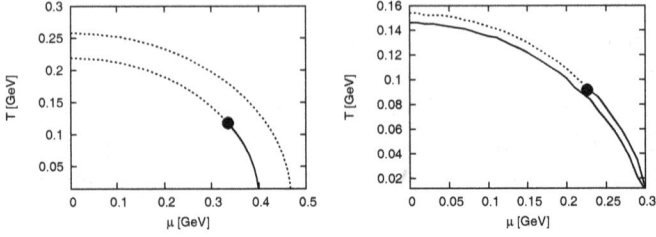

Fig. 1. (Left) Phase diagram for Case I. The upper phase boundary line corresponds to scenario 2 and the lower one corresponds to sceanrio 1. (Right) Phase diagram for Case III. The upper phase boundary is for $g_2 > 0$ and the lower one is for $g_2 < 0$. For both the cases, the solid line indicates first order phase transition. Whereas, the dashed line stands for second order phase transition for Case I and for case III it indicates crossover transition. The bold circle indicates location of the critical end point. See text for details

values of the individual parameters. Depending on the values of the various parameters we have different possible scenarios and the important results coming from each scenarios are discussed below.

Case-I: In this simple case we take the values of λ_2, g_2, $k, h = 0$. Here we have two possiblities depending on the nature of the lowest isoscalars. scenario 1 corresponds to the lowest isoscalar being quarkonium dominated. Whereas, for scenario 2, we have lowest isoscalar as tetraquark dominated meson.

The resultant phase diagram is given in Fig.1 (left). We find, for both cases for all values of the chemical potential the transition temperatures calculated from the behavior of σ and χ are the same. If we compare the phase boundaries for scenarios 1 and 2, we see for scenario 2, the order of the phase transition is second order both for low as well as high values of the chemical potential. But for scenario 1, the second order phase transition changes to weak first order phase transition above some critical value of the chemical potential, thus indicating the presence of a critical point ($T_c = 117.7$, $\mu_c = 335$ MeV).

For **Case II** we consider, $g_2 = 0$ but λ_2, $k, h \neq 0$, whereas for **Case III** g_2, λ_2, $k, h \neq 0$. But for Case III, we can further have two possiblities depending on whether g_2 is positive or negative. For $g_2 = 0$ and $g_2 > 0$ we have qualitatively same feature and thus we have included the phase boundary for Case III only in Fig. 1 (right). Same with Case I, for Cases II and III also, the transition temperatures calculated from the behavior of σ, χ are the same. As a result, the resultant phase diagram shown in Fig. 1 (right) is represented with a single phase boundary line for each case. For

$g_2 > 0$, we have a crossover transition at low chemical potential which turns into first order above some critical value of the chemical potential. Thus, we have a Critical End Point (CEP) for $g_2 = 0$ and $g_2 > 0$. The location of the critical end point for $g_2 = 0$ is ($\mu = 0.26$ GeV, $T = 0.069$ GeV) and for $g_2 > 0$ is ($\mu = 0.226$ GeV, $T = 0.0915$ GeV). Whereas, for $g_2 < 0$ we have only first order phase transition for all values of the chemical potential line owing to strong cubic interaction term for the tetraquark field. Comparing the transition temperatures for $g_2 > 0$ and $g_2 < 0$, we find the transition temperature for the former scenario is higher than the later one.

4. Discussion

In the framework of two flavor quark-meson model, we have investigated the effect of mixing between quarkonium and tetraquark fields on chiral phase transition. Depnding on the values of different coupling constants we have considered different scenarios. The common feature among all of them is that the transition for quarkonium and tetraquark happening at the same temperature for all values of the chemical potential. We find depending on the cubic self interaction term (with coupling constant g_2) of the tetraquark field, the nature of the chiral phase transition changes. While phase diagrams for $g_2 \geq 0$ have a CEP, the sufficiently strong and negative g_2 can make CEP to disappear. The transition in the later case remains always first order irrespective of high or low values of the chemical potential.

Acknowledgments

This work is supported by the NSFC under Grants No. 11250110058 and No. 11275213, DFG and NSFC (CRC 110), CAS fellowship for young foreign scientists under Grant No. 2011Y2JB05, CAS key project KJCX2-EW-N01, K.C.Wong Education Foundation and Youth Innovation Promotion Association of CAS.

References

1. A. Heinz, S. Struber, F. Giacosa, D. H. Rischke, Phys. Rev. **D 79**, 037502 (2009).
2. M. Harada, C. Sasaki, S. Takemoto. Phys. Rev. **D 81**, 016009 (2010).
3. T. K. Mukherjee and M. Huang, arXiv:1311.1313
4. O. Scavenius, A. Mocsy, I. N. Mishustin and D. H. Rischke, Phys. Rev. **C 64**, 045202 (2001).
5. B. J. Schaefer, J. Wambach, Phys. Rev. **D 75**, 085015 (2007).

Double-Λ hypernuclei at J-PARC
— E07 experiment —

K. Nakazawa* and J. Yoshida

Physics Department, Gifu University,
Gifu, Gifu 501-1193, Japan
*E-mail: nakazawa@gifu-u.ac.jp

Nuclear emulsion was used to search for double-Λ hypernuclei for a half century. We have nine events showing sequential decay of two units of Λ hyperon in nucleus, however our knowledge of Λ-Λ interaction is still poor, so far. Based on knowhow given by the past emulsion experiments, the E07 experiment at J-PARC is expected to present the events in $\sim 10^2$ and $\sim 10^3$ with *Hybrid-emulsion* and *Overall-scanning* method, respectively.

Keywords: Double-Λ hypernucleus; nuclear emulsion; hybrid-emulsion; overall-scanning.

1. Introduction

Ordinal nuclear chart shows two dimensional world of proton and neutron. However, we have been working to append the third axis of the strangeness using K^- beam at KEK with hybrid-emulsion method for a quarter century. In particular, we have presented eight events of double-Λ hyerpernuclei in nine ones for those period.[1] Although Λ-Λ interaction becomes one of key issues to understand not only Baryon-Baryon interaction under the SU(3)$_f$ symmetry, but also the Equation of State (EOS) for neutron star,[2] the information about the Λ-Λ interaction is very limited, so far.

In such condition, the E07 experiment has been approved to search for the nuclear mass dependence of Λ-Λ interaction energy, systematically, with a new hybrid-emulsion method, at J-PARC on 2006. We expect detection of $\sim 10^2$ double-Λ hypernuclei with **Hybrid-emulsion method** under the support of electric detectors' information. It necessary for the development of fully-automated system to follow tracks of Ξ^- hyperon, which shall be captured at rest by nucleus in the emulsion.

In the hybrid method, the detection yield is strongly depend on the tracking efficiency of scattered K^+ mesons by spectrometer and TOF.

Therefore, we scan all of area of the emulsion free from tagged K^+ data, so-called **Overall-Scanning method**. This method is expected to present $\sim 10^3$ samples of double-Λ hypernuclei within a few years.

2. Hybrid-emulsion Method

At the hadron hall in J-PARC, it will be provided nearly 70% pure K^- beam which is 3.5 times better than the beam at the KEK-PS. We mounted 3 times' volume of the emulsion of the previous E373 experiment, therefore it is expected for 10 times' double-Λ hypernuclei ($\sim 10^2$) to be detected.

The experimental setup of E07 is shown in Fig.1. K^- mesons with momentum of 1.66 GeV/c are irradiated to the diamond target, which cause quasi-free (K^-, K^+) reactions. Tagged events by K^+ spectrometer with KURAMA Magnet are searched for Ξ^- hyperon at SSDs located up- and down-stream of the emulsion. When a candidate track of Ξ^- hyperon is detected in the up-stream SSD without the down one, the reconstructed candidate track shall be followed down into the emulsion stack. The emulsion stack consists of 2 thin palates and 12 thick plates as shown in Fig.1. After the detection of at rest stopping of the Ξ^- hyperon by following it in plate by plate, we check X ray recorded in Ge (Hyperball) detector to study Ξ-N strong interaction by measurement of energy shift from the electromagnetic level.

Regarding the track following, fully automated system is the key technology to detect $\sim 10^2$ double-Λ hypernuclei, which is 10 times large statistics of the previous E373 spending several years. To make sure the detection within a few years, we have developed automated detection method for the emulsion surfaces, because tracks are recorded inside the emulsion. Their

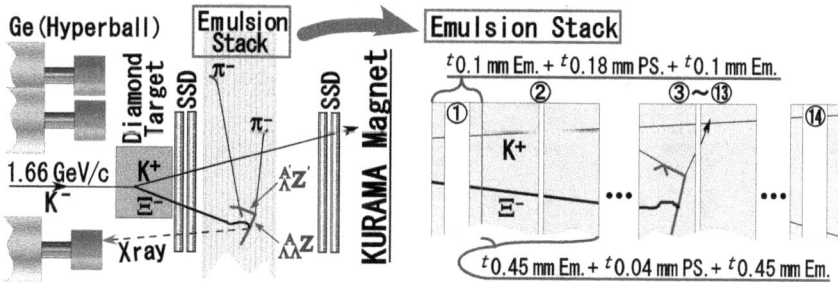

Fig. 1.   Experimental setup of the E07 experiment. The left hand shows apparatuses around the target. The composition of the emulsion stack is presented in the right.

detection accuracy achieved 2.6 ± 2.0 μm perpendicular to the emulsion plate. Position alignment between plates was the next issue, because mis-following has been easily occurred under 20 μm accuracy on E373. We succeeded to get alignment accuracy with 1.4 ± 0.8 μm in full area of 24.5 $\times$ 25 cm^2 using beam tracks recorded perpendicular to the plate. Track following inside the emulsion, the microscope stage is driven so as to set the track in the center of the field. Taking advantage of image processing technology, the fully automated track following mentioned above realizes so fast detection of double-Λ hypernuclei within a few years.

3. Overall-Scanning Method

In the hybrid method, tagging efficiency is less than 30% due to detector acceptance, tracking efficiency, and so on. It becomes loss of 70% events by Q.F. (K^-, K^+) reaction. Beside that, there are several times' events from Ξ^- hyperon originated by $'n'(K^-, K^0)\Xi^-$ reaction, where all of K^0 meson are not detected in our setup. Therefore it can be expected for so many Ξ^- stopping events in the emulsion and its number is estimated to be 10 times of the hybrid-method case. If we scan the emulsion in whole area and are able to detect typical topologies of production and decay of double-Λ hypernucleus, e.g. three vertices, $\sim 10^3$ events can be obtained.

Overall-scanning method makes it possible.[3] The method mainly consists of 2 steps which are exhaustively image taking and vertex detection via image recognition. At first, a computer-controlled optical microscope scans emulsion layers and a high-speed and -resolution camera takes their microscope images. Then, a dedicated image process picks out hypernucleus like shapes which have characteristic topology of multi tracks and vertices.

One of the key technologies is a high speed 3-dimensional scanning. Under a microscope view, a volume in an emulsion layer is projected to a 2-dimensional image. On sweeping the view in vertical and horizontal direction, 3-dimensional track information is taken as a series of cross sectional

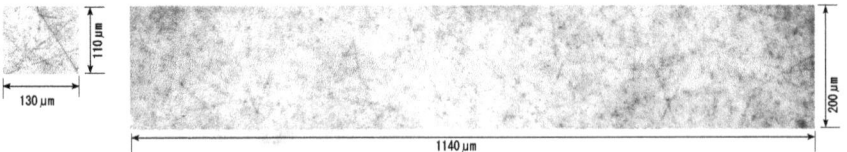

Fig. 2. Comparison of field of view of microscopic image. Image of the previous system under a $\times$50 objective (left) and the current system under a $\times$20 objective lens (right).

Table 1. Specifications of scanning systems.

System	Lens Mag.	Field [μm]	Sensor [pixel]	Frame rate [fps]	Driver	Vibration [Hz]	Driving range
Current	×20	1140×200	2048×358	800	Piezo.	5.0	500 μm
Previous	×50	130×110	512×440	100	Motor	0.18	300 μm

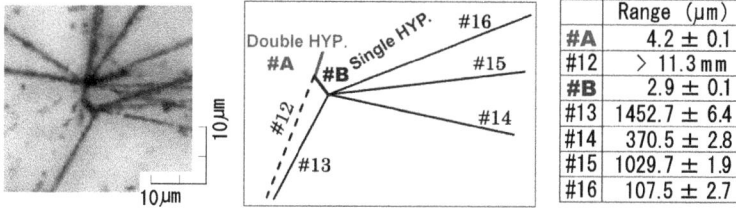

	Range (μm)
#A	4.2 ± 0.1
#12	> 11.3 mm
#B	2.9 ± 0.1
#13	1452.7 ± 6.4
#14	370.5 ± 2.8
#15	1029.7 ± 1.9
#16	107.5 ± 2.7

Fig. 3. A beam produced a double-Λ hypernucleus (#A). The particle #A decayed into a charged particle (#12) and a single-Λ which also decayed into four charged particles (#13, #14, #15, #16). A schematic drawing and the data of track length are presented.

images. We introduced some new devices for fast scanning. A high-speed and -resolution CMOS camera enables use of a lower magnitude objective lens (×20) for large field scanning as shown in Fig.2, where capture area becomes nearly 20 times wider than that obtained by the previous system. Piezoelectric actuators vibrate the lens rapidly and allows a continuous stage drive to save damping time of vibration. In total, this system is working in 800 times faster than the previous one as listed in Table 1. Among one million images, we succeeded to detect a double-Λ hypernucleus with very clear three vertices as shown in Fig.3.

4. Summary

We will study the Λ-Λ interaction via the detection of double-Λ hypernuclei in the E07 experiment at J-PARC. The development of a new *Hybrid-emulsion method* is almost finished to search for $\sim 10^2$ double-Λ hyepernuclei. After that, the *Overall-Scanning method* can be applied for the emulsion of E07, where we expect the detection of $\sim 10^3$ double-Λ hypernuclei.

Acknowledgment

This work is supported by JSPS Grant No. 23224006.

References

1. S. Aoki *et al.*, *Prog. Theor. Phys* **85** 1287 (1991); S. Aoki *et al.*, *Nucl. Phys. A* **644** 365 (1998); S. Aoki *et al.*, *Nucl. Phys. A* **828** 191 (2009); H. Takahashi *et al.*, *Phys. Rev. Lett* **87** 212502 (2001); K. Nakazawa and H. Takahashi, *Prog. Theor. Phys. Suppl.* **185** 335 (2010); J. K. Ahn *et al.*, *Phys. Rev. C* **88** 014003 (2013).
2. J. S. Bielich *Nucl. Phys. A* **804** 309 (2008).
3. J. Yoshida et al., *JPS Conf. Proc.* (in press) (2014).

Regge trajectory of the $f_0(500)$ resonance from a dispersive connection to its pole

J. Nebreda[*,1,2,3,4], J. T. Londergan[2,3], J. R. Pelaez[4] and A. P. Szczepaniak[2,3,5]

[1] Yukawa Institute for Theoretical Physics, Kyoto University, Kyoto, 606-8502, Japan
[2] CEEM, Indiana University, Bloomington, IN 47403, USA
[3] Physics Department Indiana University, Bloomington, IN 47405, USA
[4] Dpto. de Física Teórica II, Universidad Complutense de Madrid, 28040, Spain
[5] Jefferson Laboratory, 12000 Jefferson Avenue, Newport News, VA 23606, USA
[*] E-mail: jnebreda@yukawa.kyoto-u.ac.jp

We report here our results on how to obtain the Regge trajectory of a resonance from its pole in a scattering process by imposing analytic constraints in
the complex angular momentum plane. The method, suited for resonances that
dominate an elastic scattering amplitude, has been applied to the $\rho(770)$ and
the $f_0(500)$ resonances. Whereas for the former we obtain a linear Regge trajectory, characteristic of ordinary quark-antiquark states, for the latter we find
a non-linear trajectory with a much smaller slope at the resonance mass. This
provides a strong indication of the non-ordinary nature of the sigma meson.

Keywords: Regge theory; light scalar mesons.

1. Introduction

We make use of the analytical properties of the amplitudes in the complex angular momentum plane to study the Regge trajectories that link
resonances of different spins. The form of these trajectories can be used
to discriminate between the underlying QCD mechanisms responsible for
generating the resonances. In particular, linear (J, M^2) trajectories relating
the angular momentum J and the mass squared are intuitively interpreted
in terms of the rotation of the flux tube connecting a quark and an antiquark. Strong deviations from this linear behavior would suggest a rather
different nature.

Here we study the trajectory of the lightest resonances in elastic $\pi\pi$
scattering: the $\rho(770)$, which suits well the ordinary meson picture, and the
$f_0(500)$ or σ meson, whose nature is still controverted and which does not
accomodate well in the (J, M^2) trajectories.[1]

2. Regge Trajectory from a Resonance Pole

An elastic $\pi\pi$ partial wave near a Regge pole reads

$$t_l(s) = \beta(s)/(l - \alpha(s)) + f(l, s), \tag{1}$$

where $f(l, s)$ is a regular function of l, and the Regge trajectory $\alpha(s)$ and residue $\beta(s)$ are analytic functions, the former having a cut along the real axis for $s > 4m_\pi^2$.

Making use of the analyticity properties of $\alpha(s)$ and $\beta(s)$, and imposing the elastic unitarity condition, we can write down a coupled system of dispersion relations:[2]

$$\text{Re}\,\alpha(s) = \alpha_0 + \alpha' s + \frac{s}{\pi} PV \int_{4m_\pi^2}^{\infty} ds' \frac{\text{Im}\,\alpha(s')}{s'(s' - s)}, \tag{2}$$

$$\text{Im}\,\alpha(s) = \frac{\rho(s)b_0 \hat{s}^{\alpha_0 + \alpha' s}}{|\Gamma(\alpha(s) + \frac{3}{2})|} \exp\left(-\alpha' s[1 - \log(\alpha' s_0)]\right.$$

$$\left. + \frac{s}{\pi} PV \int_{4m_\pi^2}^{\infty} ds' \frac{\text{Im}\,\alpha(s') \log\frac{\hat{s}}{\hat{s}'} + \arg\Gamma\left(\alpha(s') + \frac{3}{2}\right)}{s'(s' - s)}\right), \tag{3}$$

where PV denotes "principal value" and α_0, α' and b_0 are free parameters that need to be determined phenomenologically. For the σ-meson, we modify $\beta(s)$ slightly in order to include also the Adler-zero required by chiral symmetry. In that case b_0 will not be dimensionless.

3. $\rho(770)$ and $f_0(500)$ Trajectories

For a given set of α_0, α' and b_0 parameters we solve the system of Eqs. (2) and (3) iteratively. We fix the value of the parameters by a fitting procedure in which we use only three inputs, namely, the real and imaginary parts of the resonance pole position s_M and the absolute value of the residue $|g_M|$. We require that at the pole, on the second Riemann sheet, $\beta_M(s)/(l - \alpha_M(s)) \to |g_M^2|/(s - s_M)$, with $l = 0, 1$ for $M = \sigma, \rho$. The pole parameters are taken from a precise dispersive representation of $\pi\pi$ scattering data.[3] In Fig. 1 we compare the partial waves that give the input poles[3] with the obtained Regge amplitudes on the real axis. They do not need to overlap since they are only constrained to agree at the resonance pole. Nevertheless, we find a fair agreement in the resonant region. As expected, it deteriorates as we approach threshold or the inelastic region, specially in the case of the S-wave due to the interference with the $f_0(980)$.

We show in the left panel of Fig. 2 the resulting Regge trajectories, whose parameters are given in Table 1. The imaginary part of $\alpha_\rho(s)$ is much

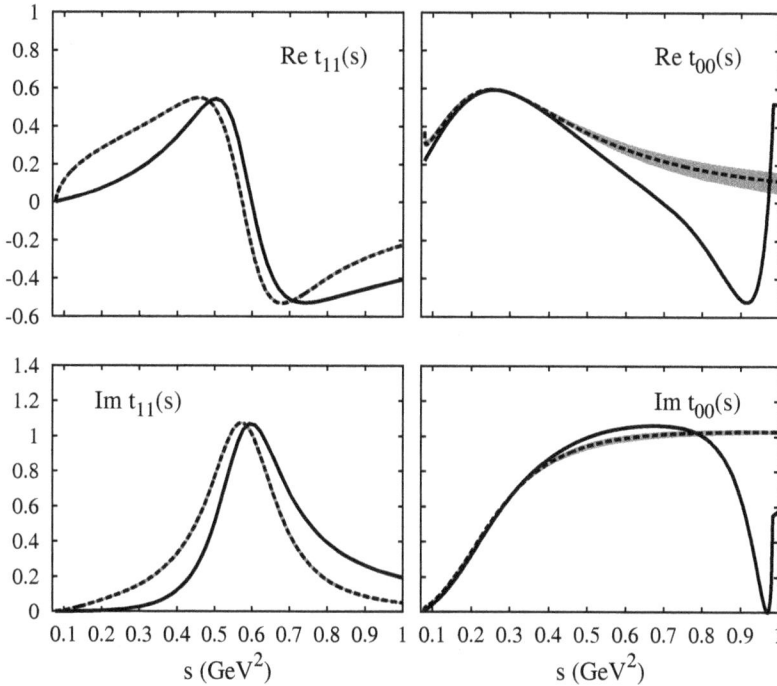

Fig. 1. Partial waves t_{lI} with $l = 1$ (left panels) and $l = 0$ (right panels). Solid lines represent the amplitudes from[3]. Their corresponding resonance poles are the input for the constrained Regge-pole amplitudes shown with dashed curves. The gray bands cover the uncertainties due to the errors in the inputs.

smaller than the real part, and the latter grows linearly with s. Taking into account our approximations, and that our error bands only reflect the uncertainty in the input pole parameters, the agreement with previous determinations is remarkable: $\alpha_\rho(0) = 0.52 \pm 0.02$,[4] $\alpha_\rho(0) = 0.450 \pm 0.005$,[5] $\alpha'_\rho \simeq 0.83\,\mathrm{GeV}^{-2}$,[1] $\alpha'_\rho = 0.9\,\mathrm{GeV}^{-2}$,[4] or $\alpha'_\rho \simeq 0.87 \pm 0.06\mathrm{GeV}^{-2}$,[6]

Table 1. Parameters of the $\rho(770)$ and $f_0(500)$ trajectories

	α_0	α' (GeV^{-2})	b_0
$\rho(770)$	0.520 ± 0.002	0.902 ± 0.004	0.52
$f_0(500)$	$-0.090^{+0.004}_{-0.012}$	$0.002^{+0.050}_{-0.001}$	$0.12\,\mathrm{GeV}^{-2}$

However, the $f_0(500)$ trajectory is evidently nonlinear and its slope is about two orders of magnitude smaller than that of the typical to quark-antiquark resonances, e.g., ρ, a_2, f_2 or π_2. This provides strong support for

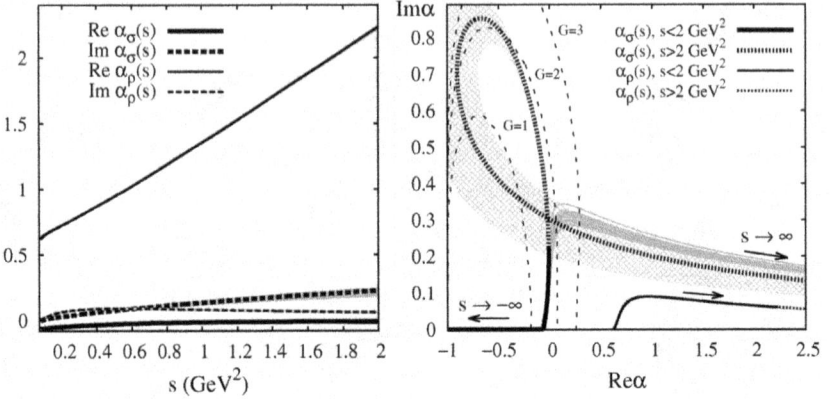

Fig. 2. (Left) $\alpha_\rho(s)$ and $\alpha_\sigma(s)$ Regge trajectories, from our constrained Regge-pole amplitudes. (Right) $\alpha_\sigma(s)$ and $\alpha_\rho(s)$ in the complex plane. At low and intermediate energies, the trajectory of the σ is similar to those of the Yukawa potential $V(r) = -Ga \exp(-r/a)/r$.[7] For the G=2 curve we can estimate $a \simeq 0.5\,\mathrm{GeV}^{-1}$, following.[7] This could be compared, for instance, to the S-wave $\pi\pi$ scattering length $\simeq 1.6\,\mathrm{GeV}^{-1}$.

a non-ordinary nature of the σ meson. Moreover, the tiny slope excludes the possibility that any of the known isoscalar resonances lie on its trajectory.

Furthermore, in Fig. 2 we show the striking similarities between the $f_0(500)$ trajectory and those of Yukawa potentials in non-relativistic scattering. Of course, our results are most reliable at low energies (thick dashed-dotted line) and the extrapolation should be interpreted cautiously. Nevertheless, our results suggest that the $f_0(500)$ looks more like a low-energy resonance of a short range potential, e.g. between pions, than a bound state of a long range confining force between a quark and an antiquark.

References

1. A. V. Anisovich, et al. Phys. Rev. D **62**, 051502 (2000).
2. G. Epstein and P. Kaus, Phys. Rev. **166**, 1633 (1968); S. -Y. Chu, et al. Phys. Rev. **175**, 2098 (1968).
3. R. Garcia-Martin, et al. Phys. Rev. Lett. **107**, 072001 (2011); R. Garcia-Martin, et al. Phys. Rev. D **83**, 074004 (2011).
4. J. R. Pelaez and F. J. Yndurain, Phys. Rev. D **69**, 114001 (2004).
5. J. Beringer, et al. (Particle Data Group), Phys. Rev. **D86**, 010001 (2012). J. R. Pelaez, PoS ConfinementX , 019 (2012) [arXiv:1301.4431 [hep-ph]].
6. P. Masjuan, et al. , Phys. Rev. D **85**, 094006 (2012).
7. C. Lovelace and D. Masson, Nuovo Cimento **26** 472 (1962). A. O. Barut and F. Calogero, Phys. Rev. **128**, 1383 (1962). A. Ahamadzadeh et al. , Phys. Rev. **131** 1315 (1963).

Recent developments on LQCD studies of nuclear force

H. Nemura

(for HAL QCD Collaboration)

Center of Computational Sciences, University of Tsukuba,
Tsukuba, Ibaraki, 305-8577, Japan
E-mail: nemura.hidekatsu.gb@u.tsukuba.ac.jp

Recent studies of (hyperonic) nuclear forces based on lattice QCD are presented. An energy independent non-local potential can be obtained from Nambu-Bethe-Salpeter wave function which is measured from the lattice QCD calculation with a set of interpolating fields of baryons, where a time-dependent Schrödinger equation in imaginary time is employed so as to avoid the exact ground state saturation in the numerical computation. The central and the tensor potentials of ΛN and ΣN with isospin $I = 2/3$ channels are presented as specific numerical examples. We also present an attempt to study the four-nucleon bound state problem using the lattice nuclear potential with stochastic variational method.

Keywords: Lattice QCD; hyperon-nucleon interaction; few-body problem.

1. Introduction

Study of nuclear force from the fundamental point of view is one of the most important and challenging problems in physics. The description of nuclei in terms of nucleonic degrees of freedom provides successful results though quantum chromodynamics (QCD) is the theory of the strong interaction and nucleons are not true fundamental building blocks of atomic nuclei but compositions of quarks and gluons. For example, modern nucleon-nucleon (NN) potentials,[1] which gives better description of NN scattering data at low energies as well as the deuteron properties, together with three-nucleon forces have been used for precise calculations of light nuclei.[2] In contrast to the normal nuclear force, phenomenological description of hyperon-nucleon (YN) and hyperon-hyperon (YY) interactions are not well constrained from experimental data, which are fundamental inputs to study the properties of hypernuclei and the hyperonic matter inside the neutron stars, since the scattering experiments are either difficult or impossible due to the short life-time of hyperons.

Thanks to both the upgrading of computer performance and various inventions of numerical algorithms, the lattice QCD approach to nuclear forces is being developed as a first-principle calculation. Not only the NN[3,4] and the YN systems[5,6] but also the light nuclei[7] and light hypernuclei[8] are the playground for the state-of-the-art lattice QCD simulations. Recently, in addition to those innovative works, a new approach to the NN interaction from the lattice QCD has been proposed.[9] In this approach, the nucleonic potential can be obtained from lattice QCD by measuring the Nambu-Bethe-Salpeter (NBS) amplitude and the observables such as the phase shifts and the binding energies are calculated through the resultant potential.[10] Especially, it seems an appealing method for studying the hyperonic forces.

In this paper, we present the central and the tensor potentials of the ΛN and the ΣN systems from lattice QCD as numerical examples of the HAL QCD method. See Ref. 10 for details and/or various applications of the method. The potential is an useful intermediate object from the lattice QCD calculation to study the strongly interacting hadronic systems in detail while the potential itself is not a physical observable. We also present a preliminary calculation of the stochastic variational method for a four-nucleon bound state problem with using a lattice NN potential which comprises central and tensor potentials defined in the even partial waves.

2. HAL QCD Approach to Interparticle Potential from Lattice QCD

In the HAL QCD scheme,[10] the non-local but energy-independent potential is defined by the Schrödinger equation as

$$(\vec{\nabla}^2 + k^2)\, \phi(\vec{r}) = 2\mu \int d^3r' U(\vec{r}, \vec{r}')\phi(\vec{r}'), \tag{1}$$

where $\phi(\vec{r})$ is the equal-time NBS wave function of two-baryon system (B_1, B_2), $\mu = m_{B_1} m_{B_2}/(m_{B_1} + m_{B_2})$ and k^2 are the reduced mass of the system and the square of asymptotic momentum in the center-of-mass frame, respectively. In practice, the nonlocal potential is expanded in terms of the velocity(derivative) as, $U(\vec{r}, \vec{r}') = V_{B_1 B_2}(\vec{r}, \vec{\nabla})\delta(\vec{r} - \vec{r}')$. The potential V may have an antisymmetric spin-orbit term when two baryons are not identical. For example, the ΛN potential is given by $V_{\Lambda N} = V_0(r) + V_\sigma(r)(\vec{\sigma}_\Lambda \cdot \vec{\sigma}_N) + V_T(r)S_{12} + V_{LS}(r)(\vec{L} \cdot \vec{S}_+) + V_{ALS}(r)(\vec{L} \cdot \vec{S}_-) + O(\nabla^2),$

where $S_{12} = 3(\vec{\sigma}_\Lambda \cdot \hat{r})(\vec{\sigma}_N \cdot \hat{r}) - \vec{\sigma}_\Lambda \cdot \vec{\sigma}_N$ is the tensor operator with $\hat{r} = \vec{r}/|\vec{r}|$, $\vec{S}_\pm = (\vec{\sigma}_N \pm \vec{\sigma}_\Lambda)/2$ are symmetric $(+)$ and antisymmetric $(-)$ spin operators, $\vec{L} = -i\vec{r} \times \vec{\nabla}$ is the orbital angular momentum operator. In the velocity expansion, $V_{0,\sigma,T}$ are the leading order (LO) potentials, denoted by V_{LO}, while $V_{LS,ALS}$ are of the next-to-leading-order (NLO). In this report, we consider the LO potentials only.

In lattice QCD simulations we first calculate the normalized four-point correlation function defined by

$$
\mathcal{R}_{\alpha\beta}^{(J,M)}(\vec{r}, t - t_0) = \sum_{\vec{X}} \left\langle 0 \left| B_{1,\alpha}(\vec{X} + \vec{r}, t) B_{2,\beta}(\vec{X}, t) \overline{\mathcal{J}_{B_3 B_4}^{(J,M)}(t_0)} \right| 0 \right\rangle \\
/ \exp\{-(m_{B_1} + m_{B_2})(t - t_0)\},
\tag{2}
$$

where the summation over $\vec{X}$ selects states with zero total momentum. The $B_{1,\alpha}(x)$ and $B_{2,\beta}(y)$ denote the interpolating fields of the baryons such as

$$
\begin{aligned}
p &= \varepsilon_{abc}(u_a C\gamma_5 d_b)u_c, \qquad \Sigma^+ = -\varepsilon_{abc}(u_a C\gamma_5 s_b)u_c, \\
\Lambda &= \tfrac{1}{\sqrt{6}}\varepsilon_{abc}\{(d_a C\gamma_5 s_b)u_c + (s_a C\gamma_5 u_b)d_c - 2(u_a C\gamma_5 d_b)s_c\},
\end{aligned}
\tag{3}
$$

and $\overline{\mathcal{J}_{B_3 B_4}^{(J,M)}(t_0)} = \sum_{\alpha'\beta'} P_{\alpha'\beta'}^{(J,M)} \overline{B_{3,\alpha'}(t_0) B_{4,\beta'}(t_0)}$ is a source operator which creates $B_3 B_4$ states with the total angular momentum J, M. This normalized four-point function can be expressed as

$$
\mathcal{R}_{\alpha\beta}^{(J,M)}(\vec{r}, t - t_0) = \sum_n A_n \sum_{\vec{X}} \left\langle 0 \left| B_{1,\alpha}(\vec{X} + \vec{r}, t) B_{2,\beta}(\vec{X}, t) \right| E_n \right\rangle \\
\times e^{-(E_n - m_{B_1} - m_{B_2})(t - t_0)},
\tag{4}
$$

where E_n $(|E_n\rangle)$ is the eigen-energy (eigen-state) of the six-quark system with the particular quantum number (i.e., J^π, M, strangeness S and isospin I), and $A_n = \sum_{\alpha'\beta'} P_{\alpha'\beta'}^{(JM)} \langle E_n | \overline{B}_{4,\beta'} \overline{B}_{3,\alpha'} | 0 \rangle$.

Since $E_n - m_{B_1} - m_{B_2} = k^2/(2\mu) + O(k^4)$, we have[11]

$$
\left(\frac{\nabla^2}{2\mu} - \frac{\partial}{\partial t}\right)\mathcal{R}(\vec{r}, t) = \int d^3 r' U(\vec{r}, \vec{r}')\mathcal{R}(\vec{r}', t) + O(k^4) = V_{LO}(\vec{r})\mathcal{R}(\vec{r}, t) + \cdots,
\tag{5}
$$

where $t - t_0$ should be moderately large so that states with large k^2 and states with more than 2 particles are suppressed.

For the spin singlet state, we extract the central potential as $V_C(r; J = 0) = (\frac{\nabla^2}{2\mu} - \frac{\partial}{\partial t})\mathcal{R}/\mathcal{R}$. For the spin triplet state, the wave function is decomposed into the S- and the D-wave components as

$$
\begin{cases}
\phi_{\alpha\beta}(\vec{r}; {}^3S_1) = \mathcal{P}\phi_{\alpha\beta}(\vec{r}; J = 1) \equiv \frac{1}{24}\sum_{\mathcal{R}\in O}\mathcal{R}\phi_{\alpha\beta}(\vec{r}; J = 1), \\
\phi_{\alpha\beta}(\vec{r}; {}^3D_1) = \mathcal{Q}\phi_{\alpha\beta}(\vec{r}; J = 1) \equiv (1 - \mathcal{P})\phi_{\alpha\beta}(\vec{r}; J = 1).
\end{cases}
\tag{6}
$$

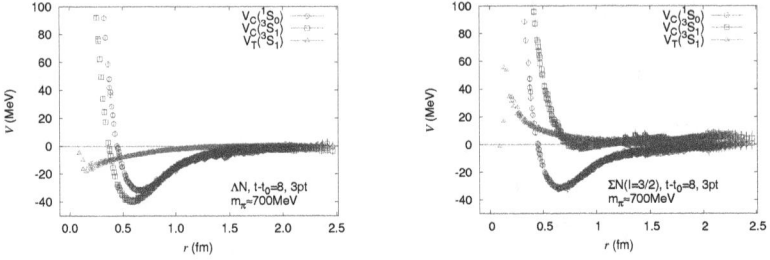

Fig. 1. Left: The central potentials in the $^1S_0$ channel (circle) and in the $^3S_1 - ^3D_1$ channel (square), and the tensor potential in the $^3S_1 - ^3D_1$ channel (triangle) of the ΛN system in $2+1$ flavor QCD as a function of r. Right: Same as the left panel but for the ΣN system with the isospin $I = 3/2$.

Therefore, the Schrödinger equation with the LO potentials for the spin triplet state becomes

$$\begin{Bmatrix} \mathcal{P} \\ \mathcal{Q} \end{Bmatrix} \times \left\{ -\frac{\nabla^2}{2\mu} + V_0(r) + V_\sigma(r)(\vec{\sigma}_\Lambda \cdot \vec{\sigma}_N) + V_T(r)S_{12} \right\} \mathcal{R}(\vec{r}, t - t_0)$$
$$= -\begin{Bmatrix} \mathcal{P} \\ \mathcal{Q} \end{Bmatrix} \times \frac{\partial}{\partial t} \mathcal{R}(\vec{r}, t - t_0).$$

(7)

The central and the tensor potentials, $V_C(r; J = 0) = V_0(r) - 3V_\sigma(r)$ for $J = 0$, $V_C(r; J = 1) = V_0(r) + V_\sigma(r)$, and $V_T(r)$ for $J = 1$, can be determined.

3. Numerical Simulations and the Results

In the present calculations, we employ $2+1$ flavor full QCD gauge configurations generated by PACS-CS collaboration[12] at $\beta = 1.9$ on a $N_L^3 \times N_T = 32^3 \times 64$ lattice. The lattice spacing at the physical quark masses is $a = 0.0907(13)$ fm,[12] thus the spatial volume is $(2.90\text{fm})^3$. We have chosen one set of hopping parameters $(\kappa_{ud}, \kappa_s) = (0.13700, 0.13640)$ for light and the strange quarks, which correspond to $(m_\pi, m_K) \cong (699, 787)$ MeV. The wall source with the Coulomb gauge fixing is placed at the time-slice t_0, and the Dirichlet boundary condition is imposed in the temporal direction at the time-slice $t - t_0 = N_T/2$. A total number of gauge configurations we have used is 399, and we put a source at $t_0 = 8n$ with $n = 0, 1, 2, \cdots, 7$ on each configuration to increase the statistics.

The central potential in the $^1S_0$ channel of the ΛN system is shown by circle in the left panel of Fig. 1. The square (triangle) symbol in the figure shows the central (tensor) potential in the $^3S_1 - ^3D_1$ channel of the ΛN system. At the distance $r \approx 0.6$ fm, the attractive well of the central potential in the $^3S_1 - ^3D_1$ channel is deeper than that of the ΛN central potential in the $^1S_0$ channel. The tensor potential itself (triangle) is weaker than the tensor potential of the NN system.[13]

The central potential in the $^1S_0$ channel of the $\Sigma N(I = 3/2)$ system is shown by circle in the right panel of Fig. 1. As seen from both panels in Fig. 1, these two central potentials in the $^1S_0$ channel, $V_C(^1S_0; \Lambda N)$ and $V_C(^1S_0; \Sigma N(I = 3/2))$, look very similar to each other, since the flavor symmetry breaking is still small in the present $2 + 1$ flavor QCD calculation, where the baryon masses are given by $(m_N, m_\Lambda, m_\Sigma) = (1.574(3), 1.635(3), 1.650(3))$ GeV.

On the other hand, the square (triangle) symbol in the right panel of Fig. 1 shows the central (tensor) potential in the $^3S_1 - ^3D_1$ channel of the $\Sigma N(I = 3/2)$ system. There is no clear attractive well in the central potential. This repulsive nature is consistent with the prediction from the naive quark model.[14] The tensor force is a little stronger than that of the ΛN system but is still weaker in magnitude than that of the NN system.

4. Stochastic Variational Calculation for Four-Nucleon Bound State

In Ref. 15, few-nucleon systems have been studied using effective central NN potentials obtained from $N_f = 3$ lattice QCD calculation. They found a four-nucleon bound state in $(L, S)J^\pi = (0, 0)0^+$ configuration with the binding energy about $B = 5.1$ MeV which can be regarded as ^{4}He at their lightest quark mass (corresponding pseudoscalar meson mass is $M_{ps} = 469$ MeV). In this section, we present stochastic variational calculation of the four-nucleon bound state with explicitly taking account of the tensor potential as well as the central potential which are obtained from the same $N_f = 3$ lattice calculation of Ref. 15.[a]

The wave function of A-body system is described by a linear combination of basis functions as

$$\Psi = \sum_{k=1}^{K} c_k \varphi_k, \quad \text{with} \quad \varphi_k = \mathcal{A}\{G(\vec{x}; A_k)[\theta_{(LL')_k}(\vec{x}; (uu')_k), \chi_{S_k}]_{JM}\eta_{kIM_I}\},$$

$$(8)$$

where c_k is the linear variational parameter determined by the variational principle, $\mathcal{A}$ is antisymmetrizer for identical particles. χ_{S_k} (η_{kIM_I}) is the spin (isospin) function of the system. $G(\vec{x}; A_k)$ is the correlated Gaussian

---

[a]See also Ref. 16 for a recent work using the lattice NN potential.

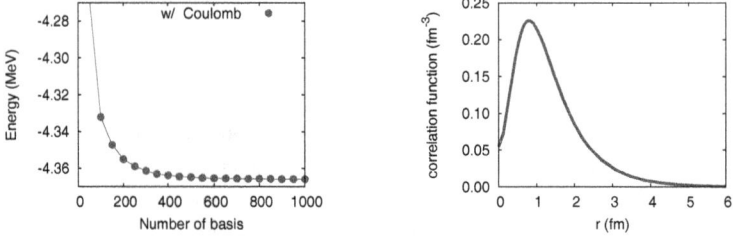

Fig. 2. Left: Energy of four-nucleon ground state as a function of the number of basis. A Serber-type lattice NN interaction is used comprising the central and the tensor potentials determined in the even partial waves. See Sec. 4 for detail. Right: Correlation function for the four-nucleon ground state obtained by using a lattice NN potential described in Sec. 4, as a function of the internucleon distance r. Coulomb potential is included.

function which is given by $G(\vec{x}; A_k) = \exp\left\{-\frac{1}{2}\sum_{i<j}^{A}\alpha_{kij}(\vec{r}_i - \vec{r}_j)^2\right\} = \exp\left\{-\frac{1}{2}\sum_{i,j=1}^{A-1}A_{kij}\vec{x}_i \cdot \vec{x}_j\right\}$. A set of relative coordinates $\{\vec{x}_1, \cdots, \vec{x}_{A-1}\}$ and the center-of-mass coordinate $\vec{x}_A$ are given by a linear transformation of single particle coordinates $\{\vec{r}_1, \cdots, \vec{r}_A\}$ such as $\vec{x}_i = \sum_{j=1}^{A}U_{ij}\vec{r}_j, (i = 1, \cdots, A)$.

In order to obtain the accurate solution of the four-nucleon bound state with explicitly utilizing the the tensor potential, we consider nonzero orbital angular momentum states $(L, S)J^{\pi} = (1, 1)0^+$ and $(2, 2)0^+$ in addition to the $(0, 0)0^+$ configuration. We employ the global vector representation[17] for these nonzero orbital angular momentum states. Therefore, the angular part of the basis function is given by

$$\theta_{(LL')_k}(\vec{x}; (uu')_k) = v_k^{L_k}v_k'^{L_k'}[Y_{L_k}(\hat{\vec{v}}_k) \times Y_{L_k'}(\hat{\vec{v}'}_k)]_{L_k},$$
$$\text{with } \begin{pmatrix} \vec{v} \\ \vec{v}' \end{pmatrix}_k = \sum_{i=1}^{A-1}\vec{x}_i\begin{pmatrix} u \\ u' \end{pmatrix}_{ki}. \tag{9}$$

Table 1. Binding energies and probabilities of total orbital angular momentum $L = 0, 1, 2$ for the four-nucleon bound state with the total angular momentum and the parity $J^{\pi} = 0^+$. The nucleon mass $m = 1161.0$ MeV is used which is observed at the lattice QCD calculation.[15]

	BE (MeV)	$P(L=0)$ (%)	$P(L=1)$ (%)	$P(L=2)$ (%)
w/ Coulomb potential	4.37	98.6	0.003	1.3
w/o Coulomb potential	5.09	98.6	0.003	1.4

The validity of the present choice of basis function is examined for several realistic NN potentials.[17] The A_{kij} and $(u, u')_{ki}$ are the nonlinear variational parameters which are determined by the stochastic variational method.[18]

At this moment, only the central and the tensor potentials determined in the even partial waves are available from the lattice QCD calculations. Thus we assume that the present lattice NN potential is a Serber-type where the potential in the odd partial wave is zero. More realistic calculation by taking account of the odd part will be performed elsewhere.

Table 1 lists the binding energies and the probabilities of orbital angular momentum state with $L = 0, 1, 2$. When we include the Coulomb potential together with the lattice NN potential the binding energy is $B = 4.37$ MeV. On the other hand, if we omit the Coulomb potential the binding energy increases a little which is $B = 5.09$ MeV. It is comparable to the result in Ref. 15 which was obtained by using the effective central NN potential. The left panel of Fig. 2 shows the energy as a function of the number of basis function, which shows relatively slower convergence than the calculation performed with the effective central potential. The tensor potential of the present lattice NN interaction is not strong so that the probability of the component with the total orbital angular momentum $L = 2$ is about 1.3 % while realistic NN potential gives the D-state probability of about or more than 10 % for ^{4}He.[17] We also calculated the correlation function which is shown in the right panel of Fig. 2.

5. Summary

We present the central and the tensor potentials of the ΛN and $\Sigma N(I = 3/2)$ systems as numerical examples of the HAL QCD method. Two potentials in the $^1S_0$ channel, one is ΛN and another is $\Sigma N(I = 3/2)$, are very similar to each other, since the SU(3) breaking effect is still small in the present calculation. Both potentials have an attractive well, so that they give attractive interactions at low energy. In the $^3S_1 - ^3D_1$ channel, while the central ΛN potential shows attraction at low energy with an attractive well, the central $\Sigma N(I = 3/2)$ potential is repulsive. Tensor potentials for both ΛN and $\Sigma N(I = 3/2)$ systems in this work are rather weak.

The stochastic variational calculation has been performed to obtain the four-nucleon bound state for a Serber-type lattice NN potential at slightly heavier quark mass corresponding the pseudoscalar meson and octet baryon masses $m_{ps} = 469$ MeV and $m_B = 1161.0$ MeV. Tensor potential as well as the central potentials determined in the even partial waves were used.

The binding energy including the Coulomb potential is $B = 4.37$ MeV, whereas it becomes $B = 5.09$ MeV if the Coulomb potential is absent. The tensor potential of the present lattice NN potential is not strong so that the D-state probability of the four-nucleon ground state is not large (about 1.3%).

Acknowledgments

The author would like to thank PACS-CS Collaboration and ILDG/JLDG[19] for allowing us to access the full QCD gauge configurations, Dr. T. Izubuchi for providing a sample FFT code, and Dr. E. Hiyama for valuable discussion on the few-body calculation. The author also thank maintainers of CPS++.[20] Calculations in this report have been performed by using the Blue Gene/L computer under the "Large scale simulation program" at KEK (No. 10-24). This research was supported in part by Strategic Program for Innovative Research (SPIRE), the MEXT Grant-in-Aid, Scientific Research on Innovative Areas (No. 25105505).

References

1. R. Machleidt and I. Slaus, *J.Phys.* **G27**, R69 (2001).
2. S. C. Pieper, *Riv.Nuovo Cim.* **31**, 709 (2008).
3. M. Fukugita, Y. Kuramashi, M. Okawa, H. Mino and A. Ukawa, *Phys.Rev.* **D52**, 3003 (1995).
4. S. Beane, P. Bedaque, K. Orginos and M. Savage, *Phys.Rev.Lett.* **97**, p. 012001 (2006).
5. S. Muroya, A. Nakamura and J. Nagata, *Nucl.Phys.Proc.Suppl.* **129**, 239 (2004).
6. S. R. Beane *et al.*, *Nucl.Phys.* **A794**, 62 (2007).
7. T. Yamazaki, Y. Kuramashi and A. Ukawa, *Phys.Rev.* **D81**, p. 111504 (2010).
8. S. Beane, E. Chang, S. Cohen, W. Detmold, H. Lin *et al.*, *Phys.Rev.* **D87**, p. 034506 (2013).
9. N. Ishii, S. Aoki and T. Hatsuda, *Phys.Rev.Lett.* **99**, p. 022001 (2007).
10. S. Aoki *et al.*, *PTEP* **2012**, p. 01A105 (2012).
11. N. Ishii *et al.*, *Phys.Lett.* **B712**, 437 (2012).
12. S. Aoki *et al.*, *Phys.Rev.* **D79**, p. 034503 (2009).
13. N. Ishii, *Few Body Syst.* **49**, 269 (2011).
14. M. Oka, K. Shimizu and K. Yazaki, *Nucl.Phys.* **A464**, p. 700 (1987).
15. T. Inoue *et al.*, *Nucl.Phys.* **A881**, 28 (2012).
16. T. Inoue *et al.*, *Phys.Rev.Lett.* **111**, p. 112503 (2013).
17. Y. Suzuki, W. Horiuchi, M. Orabi and K. Arai, *Few Body Syst.* **42**, 33 (2008).
18. Y. Suzuki and K. Varga, *Lect.Notes Phys.* **M54**, 1 (1998).
19. See http://www.lqcd.org/ildg and http://www.jldg.org.
20. Columbia physics system (cps) http://qcdoc.phys.columbia.edu/cps.html.

Spectroscopy of heavy quark hadrons

Makoto Oka

Department of Physics, H-27, Tokyo Institute of Technology
Meguro, Tokyo, 152-8551, Japan
E-mail: oka@th.phys.titech.ac.jp

Heavy quarks play special roles in the hadron spectroscopy. Some distinct features of heavy quark dynamics and their significance in the P-wave baryons with a single heavy quark are discussed. We also explore a new color configuration in exotic tetra-quark mesons with two heavy quarks. Finally, possibility of bound states of a charmed baryon with a nucleon and nuclei are examined.

Keywords: Heavy quark; heavy hadron; exotic hadron; charmed deuteron.

1. Introduction

Hadron spectroscopy has entered a new era. Recently, various new types of hadrons with heavy quarks, s, c and b, were observed in excitation spectra. The most exciting discovery was exotic "quarkonium" states, X, Y and Z's, found at Belle, BaBar and BES-III.[1] Those states cannot be explained as (quark-model) $Q\bar{Q}$ bound states, but require new dynamical contents, such as tetra-quarks, hadronic molecules and their mixings to $Q\bar{Q}$. As an example, recent analyses show that the X(3872) resonance, sitting just at the threshold of D+D* mesons, is likely to be a superposition of the $D^0\bar{D}^{*0}$, D^+D^{*-} molecules and the $c\bar{c}$ states.[2]

Such new types of hadrons may not be, in fact, new. In the light meson and baryon spectra, we have a few candidates of hadronic resonances or multi-quarks. One of them is $\Lambda(1405)$, which is likely to be dominated by the $\bar{K}N$ component.[3] Another example of tetra-quarks or meson molecules is seen in the light scalar mesons, f_0, a_0 and so on.

It is important to note that the tetra-quark and the meson-meson state are conceptually different. The former contains color-non-singlet qq components as an ingredient, while the meson molecule is composed only of color-singlet clusters. It is, however, not clear whether they are independent of each other, as they may not be distinguishable in a compact system.

In order to dissolve them we need to clarify the dynamics of quarks and multi-quark (color) correlations inside such multi-quark environments.

In this report, we investigate several elements of quark correlations and interactions in heavy quark environments. In particular, we stress that the suppression of the spin dependent interactions of heavy quarks helps us to study the light-diquark correlations.

In Sec. 2, we point out essential features of the heavy quark dynamics and symmetries. We consider the heavy baryon spectroscopy and point out possibility of exploring the diquark structures in Sec. 3. Section 4 is devoted to the doubly charmed tetra-quark states, T_{cc}, which can be produced in the e^-e^+ collision experiments. In Sec. 5, we discuss several possibilities forming charm-bound nucleus, including bound states of Λ_c with a nucleon. Conclusion is given in Sec. 6.

2. Dynamics of Heavy Quarks

The QCD lagrangian is flavor blind, i.e., the color-gauge couplings are common to all the quark flavors. It is, however, known that the scale dependence arises from the trace anomaly and Λ_{QCD} sets the typical momentum of the quark-gluon correlations. As the effective gauge coupling grows at the low energy region, the QCD vacuum has non-trivial properties, such as the quark condensates and gluon condensates.

The quark flavors can be divided into two categories, depending on their masses: the light flavors (u, d, s), and the heavy flavors (c, b, t). Spectrum of light hadrons has exhibited important roles of chiral symmetry and its spontaneous breaking (SCSB) in low-energy QCD. The SCSB tells us that the QCD vacuum has non-zero quark condensates of light flavors, $\langle \bar{u}u \rangle$, $\langle \bar{d}d \rangle$ and $\langle \bar{s}s \rangle$, and the pseudo-scalar mesons appear as the (pseudo-) Nambu-Goldstone bosons. It also induces the effective quark masses of the order of Λ_{QCD}, which leads to the constituent quark picture. The ground state mesons and baryons can be explained very well with the constituent (valence) quarks and the sum of the quark masses roughly gives the hadron mass.

In contrast, the heavy quarks are subjected to heavy-quark symmetry. In particular, the heavy quark spin symmetry plays an important role in classifying and describing the heavy hadron spectra. The dynamics behind the heavy quark symmetry is that for a large heavy quark mass, the soft QCD dynamics is independent from the mass and depends on the velocity v of the heavy quark, which is a constant of motion.[4] This symmetry gives simple relations among the transition amplitudes of heavy quark decays. Furthermore, $1/m_Q$ expansion of the QCD Hamiltonian leads to the conservation of the heavy quark spin, as is shown in the next subsection.

2.1. *Heavy quark spin symmetry (HQSS)*

The heavy-quark gluon interaction can be expanded by $1/m_Q$, where m_Q is the heavy quark mass, giving

$$\bar{Q}\gamma^\mu \frac{\lambda^a}{2} Q A_\mu^a \sim Q^\dagger \frac{\lambda^a}{2} Q A_0^a - \frac{1}{m_Q} Q^\dagger \sigma_Q \frac{\lambda^a}{2} Q \cdot (\boldsymbol{\nabla} \times \mathbf{A}^a). \tag{1}$$

The first term is the color-electric Coulomb coupling and the second term denotes the coupling of the quark spin to the color-magnetic field. One sees that the spin-dependent coupling is suppressed by $1/m_Q$. Thus in the heavy quark limit, the spin of the heavy quark disappears from the interaction. Namely, the heavy quark spin is conserved in the limit, $[\sigma_Q, H_{\text{QCD}}] = O(1/m_Q)$. This is called the heavy quark spin symmetry (HQSS).

This symmetry is quite significant in the heavy hadron spectroscopy. For instance, the pseudoscalar (0^-) meson and the vector (1^-) meson are degenerate in the limit. This is indeed the tendency for the charm and bottom meson masses, $\Delta m(\text{K}^* - \text{K}) = 397$ MeV $\rightarrow \Delta m(\text{D}^* - \text{D}) = 142$ MeV $\rightarrow \Delta m(\text{B}^* - \text{B}) = 46$ MeV. At each flavor step $(s \rightarrow c \rightarrow b)$, the splitting is reduced by a factor $\sim 1/3$.

The same reduction is manifested in the baryon spectrum, too (Fig. 1). The ud quarks in Σ_Q and Σ_Q^* have the total spin 1. Then the splittings of Σ_Q^* and Σ_Q come from the interaction between the heavy-quark spin and the light-quark spin. The observed mass splittings are 194 MeV for the strange Σ and Λ, 65 MeV for charm, and 21 MeV for bottom. Again the ratios are about $1/3$ for each step.

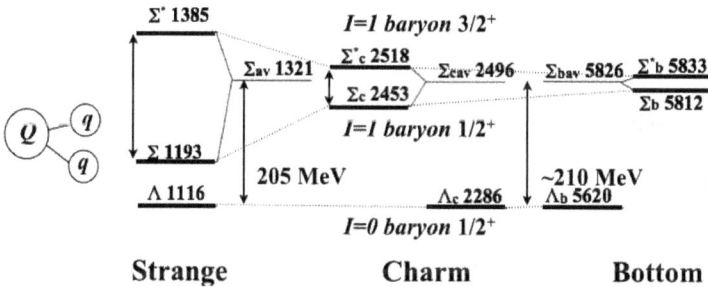

Fig. 1. Heavy baryon spectrum for strange, charm and bottom sectors.

In the heavy quark picture, the classification of the baryons are different from the familiar SU(3) scheme used in the light hadron sector. In SU(3), Σ and Σ^* belong to the octet (**8**) and the decuplet (**10**), respectively. Now in the heavy baryon spectrum, the HQSS combines Σ_Q and Σ_Q^* (**6** in SU(3)) into the same "multiplet" and Λ_Q ($\bar{\mathbf{3}}$) becomes independent from Σ_Q.

This symmetry of the heavy baryons are also viewed based on the diquark picture. The ud quarks in Λ_Q forms a spin 0 and isospin 0 (or flavor $\bar{\mathbf{3}}$) pair of light quarks. This is the scalar (0^+) diquark. The $q-q$ interaction via a gluon exchange or the instanton-light-quark coupling gives strong attraction in this channel. This interaction is about a half of the $q-\bar{q}$ interaction in the pseudoscalar (0^-) channel.

In contrast, the diquark in Σ_Q and Σ_Q^* is a spin 1, isospin 1 (or flavor **6**) diquark, called the axial-vector (1^+) diquark. The $q-q$ attraction is weaker in this channel compared to the scalar diquark. Some quenched lattice calculations are available for the diquark spectrum. Different approaches give consistent results on the mass splitting of the scalar and axial vector diquarks, ranging about 100-200 MeV.[5]

3. Heavy Baryon Spectroscopy

The quantum numbers and masses of the ground states in the heavy baryon spectrum are well understood by the quark model as well as the lattice QCD.[6,a] On the other hand, the information on excited heavy baryons is quite limited. The low-lying excitations will be P-wave baryons.

The P-wave negative-parity baryons with a single heavy quark can be classified into two distinct excitation modes: λ and ρ modes. They correspond to excitations of the individual Jacobi coordinates, λ and ρ, of the three-quark system, defined (non-relativistically) by

$$\boldsymbol{\rho} = \boldsymbol{r}_1 - \boldsymbol{r}_2$$
$$\boldsymbol{\lambda} = \boldsymbol{r}_Q - \frac{m_1 \boldsymbol{r}_1 + m_2 \boldsymbol{r}_2}{m_1 + m_2} \tag{2}$$

Here we assign the heavy quark to the particle 3, $m_3 = m_Q$. The λ-mode is the excitation from $L = 0 \to 1$ of the heavy quark relative to the light diquark, while the internal diquark state is excited in the ρ-mode.

If m_Q is equal to $m_q = m_1 = m_2$ (the SU(3) limit), then the λ and ρ modes are degenerate, and mix with each other. On the other hand, in

---

[a]One possible exception is the double charm baryon Ξ_{cc}, whose predicted masses are not completely consistent with the observed one at SELEX.[7]

the large m_Q limit, they split so that the λ mode comes lower than the ρ mode. The ratio of the excitation energies can be roughly estimated from the harmonic oscillator model of confinement,

$$\frac{E_\lambda}{E_\rho} = \frac{\omega_\lambda}{\omega_\rho} = \frac{\sqrt{m_Q + 2m_q}}{\sqrt{3m_Q}} \xrightarrow{m_Q \to \infty} \frac{1}{\sqrt{3}} \qquad (3)$$

where ω denotes the harmonic oscillator constant for the relevant mode. This ratio reaches 1 in the SU(3) limit and $1/\sqrt{3}$ in the large m_Q limit.

In reality, the ρ and λ modes may mix at intermediate m_Q. In a recent quark model calculation,[8] it is shown that the mixing is reduced rapidly when m_Q increases from $m_q \to m_s \to m_c, m_b$. Figure 2 shows the probabilities of the ρ and λ modes in the ground state Λ_Q baryon with spin 1/2. It shows a rapid change to the dominance of the λ mode when m_Q increases. At $m_Q \sim 500$ MeV, i.e., the strange constituent quark mass, the λ mode reaches 90%. Thus the heavy quark classification seems valid even in the strangeness sector.

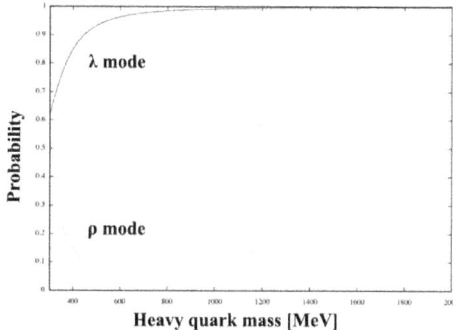

Fig. 2. Probabilities of the ρ and λ modes in the ground state of $\Lambda_Q(1/2)$ as functions of m_Q.

These two kinds of excitation modes can be generalized to higher excitations. Using the separation of these two modes, one can explore the diquark spectroscopy. As heavy quark baryons are bound states of a heavy quark and a diquark, the ρ-mode spectrum is nothing but the diquark spectrum. Furthermore, the λ modes provide us with the information of the relative motion and interaction of the diquark and the heavy quark.

Experimental studies of charm baryon spectroscopy are on-going or planned at several facilities. Among them, J-PARC has a plan to utilize the new high-momentum beam line in the hadron experimental hall to explore charm baryon excitations up to 3 GeV or higher.[9] It is also important and urgent to calculate and predict the production rates, decay widths and branching ratios of these excited baryons.

4. Production of Double Charm Exotics[b]

Another interesting object predicted in the heavy hadron spectroscopy is double heavy tetra-quark states. In particular, the double charm tetra quark, $T_{cc} = cc\bar{u}\bar{d}$ ($J^\pi = 1^+$), is a strong candidate. In organizing the $\bar{u}\bar{d}$ quarks into a scalar diquark, the T_{cc} is expected to be bound significantly from the lowest strong-decay threshold, DD*.[11,c]

As one of the possible production modes is the e^+e^- collision in the B factory, we recently calculated the production rates of T_{cc}'s in the NRQCD formulation.[10] In this work, we have shown that we have two low-lying states with $I(J^\pi) = 0(1^+)$, having different color structures. As the scalar diquark with color **3** is a favored one in QCD, the lowest mode is $T_{cc}[3, {}^3S_1]$, in which the quantum number of $\bar{u}\bar{d}$ quarks are given in the bracket. We point out that the second lowest state is $T_{cc}[\bar{6}, {}^1S_0]$, in which both (cc) and $(\bar{u}\bar{d})$ form color **6** diquark with $J^\pi = 0^+$ and 1^+, respectively. This is very interesting because such diquarks with exotic color quantum number can exist only inside multi-quark systems. Therefore observing such a state at a predicted mass will be a strong support on the assumed dynamics (of quark confinement) employed in its prediction.[d]

Belle reported productions of double charm hadrons in the e^+e^- collision experiment at the B factory, KEKB.[13] Recombination of the charm quarks will produce double charmed mesons (T_{cc}) as well as baryons (Ξ_{cc}) in similar production processes. In order to predict the production cross section, we employ the NRQCD formulation.[10] Hard processes calculated perturbatively to the leading order in α_s are factorized from soft processes, representing fragmentations to the hadron. We found that the $T_{cc}[3, {}^3S_1]$ and $T_{cc}[\bar{6}, {}^1S_0]$ show different momentum dependences of the outgoing tetra quark and therefore can be distinguished in a production experiment.

---

[b]This section is based on Ref. 10 in collaboration with T. Hyodo, Y.R. Liu, K. Sudoh and S. Yasui.

[c]Note that T_{cc} cannot decay into DD because no 1^+ state can be formed by DD. It is, however, possible for T_{cc} to decay radiatively into DD$\gamma(M1)$.

[d]These two states may mix with each other, while the mixing is suppressed in the HQ limit by factor $1/m_Q$, as the transition requires the spin flip of cc. This is numerically confirmed in a dynamical 4-quark calculation.[12]

5. Charmed Nuclei[e]

It is very interesting to study possible bound states of heavy hadrons in nuclei, because such bound states will give us precious information on the inter-baryon interactions and will open a new branch of exotic nuclear physics. As the first step, we consider the charmed deuteron, *i.e.*, bound state of a charmed baryon and a nucleon.

In recent studies,[14] we constructed a meson exchange potential model for Λ_c and N, based on an effective theory with chiral symmetry and heavy-quark symmetry. Only the light mesons are exchanged between the baryons and the coupling constants are determined by the use of the strong decays of the heavy baryons, the quark model, the chiral multiplet assumption, the vector meson dominance and the QCD sum rule results. Short-range interactions are either replaced by the cut-off form factors or the quark-model based repulsion. Details are given in Ref. 14.

We have found a significant attraction induced by channel couplings involving Σ_c and Σ_c^*. For instance, the $\Lambda_c N(J^\pi = 0^+)$ bound state is realized by coupling three channels, $\Lambda_c N(^1S_0)$, $\Sigma_c N(^1S_0)$, and $\Sigma_c^* N(^5D_0)$. Among them the last channel couples very strongly by the tensor force, which comes mainly from the one-pion exchange (OPE) between the baryons. As a result, $\Sigma^* N$ mixes significantly in the $\Lambda_c N$ bound state.

Seven channels contribute in the $\Lambda_c N(1^+)$ system; $\Lambda_c N(^3S_1)$, $\Sigma_c N(^3S_1)$, $\Sigma_c^* N(^3S_1)$, $\Lambda_c N(^3D_1)$,$\Sigma_c N(^3D_1)$, $\Sigma_c^* N(^3D_1)$, and $\Sigma_c^* N(^5D_1)$. Among them, the D-wave channels are strongly mixed with the main component of the bound state.

Our conclusions on the charmed deuteron are as follows. (1) The OPE tensor term induces strong mixings of Σ_c and Σ_c^*, which is a driving force of the $\Lambda_c N(0^+, 1^+)$ bound states. (2) Inclusion of the vector and scalar mesons gives larger binding energies for both the channels. (3) The binding energies depend sensitively to the choices of the cutoff parameters and the short-range repulsion. By adjusting the short-range repulsion to the NN force using the quark cluster model, it is found that the two-body bound states are rather shallow at most, but there may exist charmed hypernuclei with $A \geq 3$ with significant binding energies. (4) The interactions in the 0^+ and 1^+ channels are so different that they give quantitatively similar results.

Double charm dibaryon states (siblings of the H dibaryon) were also

---

[e]This section is based on Ref. 14 in collaboration with Y. R. Liu, S. Maeda, A. Yokota and E. Hiyama.

studied using the OPE potential model.[15] The results for $\Lambda_c\Lambda_c(0^+)$ are quite similar to the $\Lambda_c N$ case. We found that the effect of channel couplings to the D wave states, $\Sigma_c^*\Sigma_c^*(^5D_0)$ and $\Sigma_c\Sigma_c^*(^5D_0)$, give enough attraction to produce a bound state with a moderate binding energy.

6. Conclusion

I conclude this paper with a few short sentences. The heavy hadron spectroscopy contains very rich physical contents and high potentiality. More theoretical efforts should be devoted to analyses of heavy hadron spectra and structures. I hope that new facilities, such as Belle, BESIII, J-PARC, PANDA, will further accelerate the development in this field.

Acknowledgments

The author would like to thank the organizers of Chiral-13 for hospitality. He also thanks the collaborators of the researches presented here; Drs. T. Hyodo, Y. R. Liu, K. Sudoh, S. Yasui, A. Hosaka, E. Hiyama, A. Yokota, T. Yoshida, K. Sadato, and S. Maeda. He also acknowledges useful discussions with Drs. H. Noumi and K. Ozawa.

References

1. J. Brodzicka et al. (Belle Coll.), Prog. Theor. Exp. Phys. **2012** (2012) 04D001; N. Brambilla et al., Eur. Phys. J. **C71** (2011) 1534; J. Messchendorp et al. (BESIII Coll.), PoS **Bormio2013** (2013) 043.
2. M. Takizawa, S. Takeuchi, Prog. Theor. Exp. Phys. **2013** (2013) 0903D01.
3. T. Hyodo, D. Jido, Prog. Part. Nucl. Phys. **67** (2012) 55.
4. H. Georgi, Phys. Lett. **B240** (1990) 447.
5. M. Hess, et al., Phys. Rev. **D58** (1998) 111502; C. Alexandrou, et al., Phys. Rev. Lett. **97** (2006) 222002; R. Babich, et al., Phys. Rev. **D76** (2007) 074021; T. DeGrand, et al., Phys. Rev. bf D77 (2008) 034505.
6. Y. Namekawa et al. (PACS-CS Coll.), Phys. Rev. **D87** (2013) 094512.
7. M. Mattson et al. (SELEX Collaboration), Phys. Rev. Lett. **89** (2002) 112001.
8. T. Yoshida et al., in preparation.
9. H. Noumi, Few-Body Syst., **54** (2013) 813.
10. T. Hyodo, Y.R. Liu, K. Sudoh, S. Yasui, Phys. Lett. **B721** (2013) 56.
11. S. Zouzou, et al., Z. Phys. **C30** (1986) 457; H. J. Lipkin, Phys. Lett. **B172** (1986) 242.
12. J. Vijande, A. Valcarce, Phys. Rev. **C80** (2009) 035204.
13. K. Abe, et al., Belle Collaboration, Phys. Rev. Lett. **89** (2002) 142001.
14. Y.R. Liu and M. Oka, Phys. Rev. **D85** (2012) 014015; S. Maeda, et al., in preparation.
15. W. Meguro, Y.R. Liu, M. Oka, Phys. Lett. **B704** (2011) 547.

Recent developments on hadron interaction and dynamically generated resonances

E. Oset[†,§], M. Albaladejo[†] and Ju-Jun Xie[†,‡]

[†] *Departamento de Física Teórica and IFIC, Centro Mixto Universidad de Valencia-CSIC, Institutos de Investigación de Paterna, Aptdo. 22085, 46071 Valencia, Spain*
[‡] *Institute of Modern Physics, Chinese Academy of Sciences, Lanzhou 730000, China, and State Key Laboratory of Theoretical Physics, Institute of Theoretical Physics, Chinese Academy of Sciences, Beijing 100190, China*
[§] *E-mail: oset@ific.uv.es*

A. Ramos

Departament d'Estructura i Constituents de la Matèria and Institut de Ciències del Cosmos, Universitat de Barcelona, Martí i Franquès 1, E-08028 Barcelona, Spain

In this talk I report on the recent developments in the subject of dynamically generated resonances. In particular I discuss the $\gamma p \to K^0 \Sigma^+$ and $\gamma n \to K^0 \Sigma^0$ reactions, with a peculiar behavior around the $K^{*0}\Lambda$ threshold, due to a $1/2^-$ resonance around 2035 MeV. Similarly, I discuss a BES experiment, $J/\psi \to \eta K^{*0}\bar{K}^{*0}$ decay, which provides evidence for a new h_1 resonance around 1830 MeV that was predicted from the vector-vector interaction. A short discussion is then made about recent advances in the charm and beauty sectors.

Keywords: Hadron interaction; dynamically generated resonances.

1. Introduction

The unitary treatment of coupled channels with interaction kernels extracted from chiral Lagrangians has given rise to the chiral unitary approach that provides scattering amplitudes for hadron hadron interaction and, by looking at poles of the scattering amplitudes, bound states or resonances which we call dynamically generated.[1] The chiral Lagrangians are extrapolated to include vector mesons by means of the local hidden gauge Lagrangians[2] and then one can study vector-vector and vector-baryon interactions.[3] In the vector-vector sector this interaction has been shown to lead to many resonances,[4] some of which can be associated to known states, while a few others are predictions. One of these is a h_1 $(0^-(1^{+-}))$ state which couples only to $K^*\bar{K}^*$.

In the vector-baryon sector, several resonances around 2000 MeV have been found in Ref. 5 from the interaction of vectors with the members of the proton SU(3) octet. Some $1/2^-$ states appear from the interaction which can be associated to known states, but some are predictions. In fact, there are predictions of states around 2000 MeV that could be associated to states previously catalogued in the PDG, but which have disappeared in the latest edition of the Book.

In the charm and beauty sectors the proliferation of X, Y, Z mesonic states which do not fit the conventional charmonium spectrum have led to a plethora of works suggesting molecules, tetraquarks, and other exotic states.[6] The baryon sector has followed this trend with multiple suggestions of non conventional baryons.[7] In this talk I will address some examples in these sectors.

2. Signature of an h_1 State in the $J/\psi \to \eta K^{*0}\bar{K}^{*0}$ Decay

We plot diagrammatically the process in Fig. 1. If we produce an η and a $K^*\bar{K}^*$ in the most favorable process involving L=0 in the vertex, then the quantum numbers of the $K^*\bar{K}^*$ pair are $I^G(J^{PC}) = 0^-(1^{+-})$. These are the quantum numbers of an h_1 resonance. The produced $K^*\bar{K}^*$ will interact, which is depicted by the bubbles in the figure signifying the multiple interactions of the Bethe Salpeter equation. Then, if this interaction is strong enough to produce a resonance, as found in Ref. 4, a peak will be produced in the invariant mass of a final $K^{*0}\bar{K}^{*0}$ state. This is what is seen in the experiment,[8] as shown in Fig. 2. In Ref. 9 the evaluation is done for this invariant mass distribution and taking the input from Ref. 4, up to a an arbitrary normalization, the curves of Fig. 2 are obtained which provide a good reproduction of the data. A choice is made there of the subtraction constant of the loop function for the two $K^*\bar{K}^*$ mesons. With this subtraction constant demanded by experiment, one can go back to[4] and evaluate the $K^*\bar{K}^*$ scattering matrix with the h_1 quantum numbers, and one finds that the amplitude contains a pole corresponding to a resonance around 1830 MeV, in qualitative agreement with the prediction around 1800 MeV. The experimental data have served to obtain additional information that can make the prediction more precise. Hence the experiment provides evidence for a new h_1 state which is so far no catalogued.

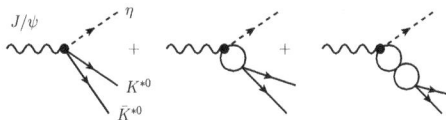

Fig. 1. Diagrammatic representation of the $J/\psi \to \eta K^{*0}\bar{K}^{*0}$ decay.

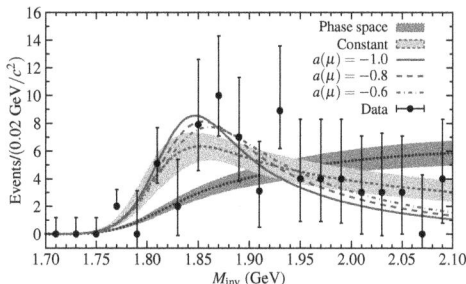

Fig. 2. (Color online) The $K^{*0}\bar{K}^{*0}$ invariant mass spectrum of $J/\psi \to \eta K^{*0}\bar{K}^{*0}$ decay. The data points are taken from Ref. 8. The different lines represent the output for different approaches. The short-dashed line and the associated error band (light blue) represent the results of a constant potential. The (red) solid, long-dashed and dot-dashed lines represent the results for the potential of[4] with $a(\mu) = -1.0$, -0.8 and -0.6, respectively. Finally, the (green) dotted line, and the associated error band (dark green) is the prediction for phase space alone.

3. The $\gamma p \to K^0\Sigma^+$ and $\gamma n \to K^0\Sigma^0$

In a recent experiment,[10] the cross section for the $\gamma p \to K^0\Sigma^+$ reaction exhibits a curious behavior close to the $K^*\Lambda$ threshold. The cross section suddenly drops and the differential cross sections become isotropic. The cross section is not reproduced by any of the standard models MAID, SAID, etc. A hint was given in Ref. 10 that this could be due to the role played by intermediate states of vector-baryon. A materialization of this idea was recently made in Ref. 11. In this work the mechanism for $K^0\Sigma^+$ photoproduction was taken as depicted in Fig. 3. The photon gets converted into a vector meson according to the vector dominance hypothesis.[2] Then the vector and the nucleon interact, as done in Ref. 5 and finally, a vector and a baryon produce the final $K^0\Sigma^+$ state via the exchange of a pion. In the energy region of interest the vector-baryon states of relevance are $K^{*+}\Lambda$, $K^{*+}\Sigma^0$ or $K^{*0}\Sigma^+$ and we take all of them into consideration. We then repeat the same with the $\gamma n \to K^0\Sigma^0$ reaction.

The cross section for the reaction shows up a peak due to a N^* dynamically generated resonance around 2000 MeV,[5] but the striking thing is that there is a destructive interference between the intermediate $K^*\Sigma$ and $K^*\Lambda$ channels such that the cross section for the proton falls down precisely where the individual contributions have the peak, as can be seen in Fig. 4. On the other hand, in the case of a neutron target, the $K^*\Lambda$ contribution is weaker and the peak of the cross section is still seen after the interference, such that the cross section on the neutron develops a peak where the cross section on the proton falls down. This is a net prediction of the theory which would be very interesting to be observed. Although with some difference, since we have more possibilities in the intermediate states, these findings are a reminiscence of those for η photoproduction on protons and neutrons found in Ref. 12, where a peak seen in the photoproduction on the neutron targets was interpreted as a consequence of the different interference of the $K\Sigma$ and $K\Lambda$ channels. In the case of γn, the intermediate state for $K\Lambda$ is $K^0\Lambda$ and the photon does not couple to the neutral K^0, hence there is only contribution from the $K\Sigma$ channel in that case.

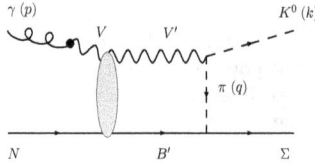

Fig. 3.   Mechanism for the photoproduction reaction $\gamma N \to K^0\Sigma$. The symbol V stands for the ρ^0, ω and ϕ mesons, while $V'B'$ denotes the intermediate channel, which can be $K^{*+}\Lambda$, $K^{*+}\Sigma^0$ or $K^{*0}\Sigma^+$, in the case of $\gamma p \to K^0\Sigma^+$, and $K^{*+}\Sigma^-$ or $K^{*0}\Lambda$, in the case of $\gamma n \to K^0\Sigma^0$, the $K^{*0}\Sigma^0$ intermediate channel does not contribute due to the zero value of the $\pi^0\Sigma^0\Sigma^0$ coupling at the Yukawa vertex.

With the input used in Ref. 5 we find the results of Fig. 5 shown by the solid line. The fall down appears but is displaced with respect to the experimental data. This serves us to fine tune the subtraction constant in the meson baryon loop function. Once this is changed, the agreement with data is then good. With the new parameters we then evaluate the cross section for the neutron target and we find a peak in the cross section at 2075 MeV, which is a prediction. Furthermore, with these parameters we can now evaluate the poles of the vector-baryon amplitude, finding one around 2030 MeV. Thus, the data can be interpreted as the effect of a dynamically generated resonance $(1/2^-, 3/2^-)$ at 2030 MeV.

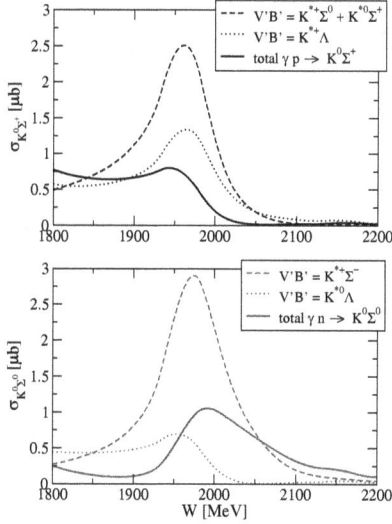

Fig. 4. Contributions to the $\gamma p \to K^0 \Sigma^+$ (upper panel) and $\gamma n \to K^0 \Sigma^0$ (lower panel) cross sections.

4. Dynamically Generated States in the Charm and Beauty Sectors

This topic has generated much activity[13–23] and recent short reviews can be seen in Refs. 6 and 7. We shall address here some recent work on the topic.

In Refs. 24 and 25 baryon states of hidden charm were investigated using an extrapolation to SU(4) of the results of the local hidden gauge in SU(3). Actually the dominant terms in the potential come from the exchange of light vectors where the heavy quarks play the role of spectators. In this case the results can be obtained from a mapping of SU(3). It has been found recently[28,29,31] that these results are consistent with heavy quark spin symmetry.[26,27]

In Refs. 25 and 24 one obtains a N^* and a Λ^* states in the pseudoscalar-baryon and vector-baryon cases. The nucleon state, with mass around 4260 MeV couples mostly to $\bar{D}\Sigma_c$, while there are two Λ, one around 4200 MeV, which couples mostly to $\bar{D}_s\Lambda_c^+$ and $\bar{D}\Xi_c$, and another one at 4400 MeV, which couples mostly to $\bar{D}\Xi_c'$. Similar states are found substituting the $\bar{D}$ by $\bar{D}^*$ displaced by about 150 MeV with respect to the PB states. These

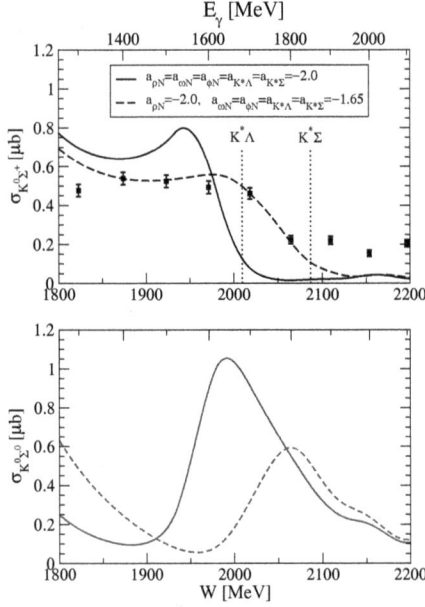

Fig. 5. Upper panel: Comparison of the $\gamma p \to K^0 \Sigma^+$ cross section, obtained with two parameter sets, with the CBELSA/TAPS data of Ref.[10] The downfall of the cross section allows one to redefine the parameters of the model and give a better prediction for the position of the resonance. Lower panel: Predictions for the $\gamma n \to K^0 \Sigma^0$ cross section using two parameter sets.

states are also found to have relatively narrow widths of about 50 MeV. It is thus quite surprising to find N and Λ states with such a large mass and a small width. The reason is that they can decay to lighter channels of PB or VB, but the transitions are penalized by forcing the exchange of a heavy vector that reduces the strength of the matrix element.

In a more recent paper, these ideas are generalized to the beauty sector[30] and again the states obtained have masses of around 11000 MeV.

In Ref. 28 baryon states of hidden charm are studied from the perspective of heavy quark spin symmetry using dynamics of the local hidden gauge. The results of Ref. 25 are reproduced and extended to other combinations of pseudoscalar-baryon or vector-baryon. In Ref. 29 this is extended to the beauty sector, and one again, the results of Ref. 30 are reproduced and more states are predicted from different combinations of meson-baryon channels. An extrapolation of these ideas is also done to the meson-meson sector, where meson states of hidden beauty are generated.[31] A fair amount of states is predicted which we hope will be found experimentally in the future.

Another source of predictions in this sector comes from the use of Heavy quark spin symmetry, using some phenomenological input, where results qualitatively similar to those reported before are also obtained.[32-34] A variety of methods are also used in other works.[35-38] The list of work done in this issue is long giving evidence of a thriving field.

5. Conclusions

The chiral unitary approach has proved very efficient to study the interaction of hadrons, and sometimes the scattering matrices show poles that are consequence of the unitarization with the potentials obtained from the chiral Lagrangians, signalling the existence of dynamically generated states, which are a kind of molecular states of some hadrons. Mounting experimental information is giving support to the nature of these states by finding some of the states predicted that had not been reported before. On the other hand, and curiously, it has been in the charm and beauty sectors that many states have been found that do not fit in the ordinary scheme of the quark model and call for more complex structures, some of them clearly of molecular nature. The theoretical effort in these sectors is impressive, as well as the amount of states predicted. With the continuing observation of new states in these sectors in the different facilities around the world, we can only hope that these predicted states are gradually found and we deepen our understanding of hadron physics.

Acknowledgments

This work is partly supported by the Spanish Ministerio de Economía y Competitividad and European FEDER funds under the contract number FIS2011-28853-C02-01 and FIS2011-28853-C02-02, and the Generalitat Valenciana in the program Prometeo, 2009/090. This work is also partly supported by the National Natural Science Foundation of China under grant 11105126. We acknowledge the support of the European Community-Research Infrastructure Integrating Activity Study of Strongly Interacting Matter (HadronPhysics3, Grant Agreement n. 283286) under the Seventh Framework Programme of the EU.

References

1. J. A. Oller, E. Oset and A. Ramos, Prog. Part. Nucl. Phys. **45**, 157 (2000) .
2. M. Bando et al., Phys. Rev. Lett. **54**, 1215 (1985).

3. E. Oset et al., Int. J. Mod. Phys. E **21**, 1230011 (2012) .

4. L. S. Geng and E. Oset, Phys. Rev. D **79**, 074009 (2009).

5. E. Oset, A. Ramos, Eur. Phys. J. **A44**, 445-454 (2010) .

6. W. Chen et al. , arXiv:1311.3763 [hep-ph].

7. B. S. Zou, Nucl. Phys. A **914**, 454 (2013) .

8. M. Ablikim *et al.* [BES Collaboration], Phys. Lett. B **685**, 27 (2010) .

9. J. -J. Xie, M. Albaladejo and E. Oset, arXiv:1306.6594. Phys. Lett. B.

10. R. Ewald *et al.*, Phys. Lett. B **713**, 180 (2012)

11. A. Ramos and E. Oset, Phys. Lett. B **727**, 287 (2013) .

12. M. Doring and K. Nakayama, Phys. Lett. B **683**, 145 (2010) .

13. E. E. Kolomeitsev and M. F. M. Lutz, Phys. Lett. B **582**, 39 (2004) .

14. J. Hofmann and M. F. M. Lutz, Nucl. Phys. A **733**, 142 (2004) .

15. F. -K. Guo et al., Phys. Lett. B **641**, 278 (2006) .

16. D. Gamermann, E. Oset, D. Strottman and M. J. Vicente Vacas, Phys. Rev. D **76**, 074016 (2007) .

17. D. Gamermann and E. Oset, Eur. Phys. J. A **33**, 119 (2007) .

18. J. Hofmann and M. F. M. Lutz, Nucl. Phys. A **763**, 90 (2005) .

19. T. Mizutani and A. Ramos, Phys. Rev. C **74**, 065201 (2006) .

20. R. Molina and E. Oset, Phys. Rev. D **80**, 114013 (2009)

21. C. Garcia-Recio et al, Phys. Rev. D **87** (2013) 074034 .

22. C. Garcia-Recio et al., Phys. Rev. D **87** (2013) 034032 .

23. O. Romanets et al., Phys. Rev. D **85** (2012) 114032 .

24. J. -J. Wu, R. Molina, E. Oset and B. S. Zou, Phys. Rev. Lett. **105**, 232001 (2010) .

25. J. -J. Wu, R. Molina, E. Oset and B. S. Zou, Phys. Rev. C **84**, 015202 (2011)

26. N. Isgur and M. B. Wise, Phys. Lett. B **232**, 113 (1989).

27. M. Neubert, Phys. Rept. **245**, 259 (1994) .

28. C. W. Xiao, J. Nieves and E. Oset, Phys. Rev. D **88**, 056012 (2013) .

29. C. W. Xiao and E. Oset, Eur. Phys. J. A **49**, 139 (2013) .

30. J. -J. Wu and B. S. Zou, Phys. Lett. B **709**, 70 (2012) .

31. A. Ozpineci, C. W. Xiao and E. Oset, Phys. Rev. D **88**, 034018 (2013) .

32. C. Hidalgo-Duque, J. Nieves and M. P. Valderrama, Phys. Rev. D **87**, 076006 (2013).

33. J. Nieves and M. P. Valderrama, Phys. Rev. D **86**, 056004 (2012).

34. F. -K. Guo, C. Hidalgo-Duque, J. Nieves and M. P. Valderrama, Phys. Rev. D **88**, 054007 (2013) .

35. Y. -R. Liu, M. Oka, M. Takizawa, X. Liu, W. -Z. Deng and S. -L. Zhu, Phys. Rev. D **82**, 014011 (2010).

36. Y. -J. Zhang, H. -C. Chiang, P. -N. Shen and B. -S. Zou, Phys. Rev. D **74**, 014013 (2006).

37. M. T. Li, W. L. Wang, Y. B. Dong and Z. Y. Zhang, Int. J. Mod. Phys. A **27**, 1250161 (2012).

38. M. Cleven, Q. Wang, F. -K. Guo, C. Hanhart, U. -G. Meiner and Q. Zhao, arXiv:1310.2190 [hep-ph].

Power counting in nuclear effective field theory

M. Pavon Valderrama

Institut de Physique Nucléaire,
Université Paris-Sud, IN2P3/CNRS,
F-91406 Orsay Cedex, France,
pavonvalderrama@ipno.in2p3.fr

The effective field theory formulation of nuclear forces is able to provide a systematic and model independent description of nuclear physics, where all processes involving nucleons and pions can be described in terms of the same set of couplings, the theoretical errors are known in advance and the connection with QCD is present. These features are a consequence of renormalization group invariance, which in turn determines the power counting of the theory. Here we present a brief outline of how to determine the power counting of nuclear effective field theory, what does it looks like and what are the predictions for the two-nucleon sector at lowest orders.

1. Introduction

The theoretical understanding of nuclear forces is still one of the most important problems of nuclear physics.[1] Nowadays we know that the fundamental theory of strong interactions is quantum chromodynamics (QCD) and hence nuclear forces should be explained in its terms. Yet QCD is only explicitly solvable at high energies, where it is perturbative. At low energies QCD becomes non-perturbative and we must rely on numerical methods – lattice QCD – to solve it. Owing to the huge computational effort involved, nuclear physics still lies beyond the range of lattice QCD, though there are a few calculations for pion masses larger than the physical one.[2,3]

There is however a more indirect way to connect nuclear physics and QCD, which is the renormalization group evolution (RGE) of QCD from high to low energy. Though the RGE equations of QCD can not be solved exactly at long distances, we know that the necessary outcome of this procedure is a low energy effective field theory (EFT) involving the fields and symmetries that survive at the typical length scales of nuclear physics. As a consequence of RGE, everything within the EFT can be organized as a

Taylor expansion in which the expansion parameter is the ratio of the light energy scales (Q) over the heavy energy scales (M);

$$\mathcal{A} = \sum_{\nu} \hat{A}^{(\nu)} \left(\frac{Q}{M}\right)^{\nu}. \tag{1}$$

This idea has been a great success in processes involving light mesons and at most one hadron, where it goes under the name of Chiral Perturbation Theory (ChPT).[4] But the application of ChPT to nuclear physics is not straightforward. Nuclear forces are non-perturbative, as the existence of deuteron demonstrates. This feature has been for a long time a serious drawback for the correct application of the EFT idea to nuclear physics: not only is ChPT a perturbative theory, but also most of the traditional EFT knowledge is perturbative as well.

For dealing with these loopholes Weinberg proposed a practical work-around.[5,6] First he made the observation that the nuclear potential admits a well-defined perturbative expansion and can be derived from ChPT. But the real problem is the EFT expansion of observable quantities. His proposal, known as the Weinberg power counting, is to do as in traditional nuclear physics: plug the EFT potential into the Schrödinger equation, obtain wave functions and finally compute observables. Actually, this amounts to ignoring the problem with observables in the hope that they will implicitly follow a power counting.

However, it turns out that observables in this scheme do not follow a power counting. The reason is that the Weinberg counting is not renormalizable, by which we mean the following: the EFT description and the fields it involves – like the nucleon and the pion – are only valid at low momenta. At high momenta the EFT amplitudes are no longer valid: a pion hitting a nucleon with a 1 GeV momentum is not an accurate physical picture. What actually happens at 1 GeV is that a few quarks and gluons collide. Thus EFTs involve a momentum cut-off Λ that is used to eliminate all the unphysical high momentum components of the theory. Yet EFTs arise from the infrared RGE of a general theory, meaning that they should be independent of the reference scale or the cut-off used to define them. That is, EFTs should be cut-off independent, or in other word *renormalizable*. Otherwise the RGE from which power counting is derived will be broken (and hence power counting). However Nogga, Timmermans and van Kolck found out[7] that the amplitudes obtained in the Weinberg counting are not cut-off independent even at leading order (neither they are at subleading orders[8,9]). Without renormalizability, the EFT predictions do not have well-defined theoretical errors and what is even worse, they lack a meaningful connection with QCD, which is the original motivation for using EFT. Thus the search for a consistent nuclear EFT had to begin all over again.

In hindsight,the construction of a consistent nuclear EFT is trivial. It just involves to follow a series of simple steps [a]:

1) Define which are the leading and subleading order diagrams that conform the EFT potential.
2) Solve the leading order potential non-perturbatively.
3) Include the subleading order corrections as perturbations.
4) Finally check for cut-off independence at each order:

   4.a) If calculations are cut-off independent, we are done. We can proceed to the next order if we want more accuracy.
   4.b) If not, we include new contact interactions until we achieve cut-off independence.

The implementation of these steps is straightforward now that the technical details about how to renormalize singular interactions have been finally understood.[8–11] The outcome is a theory in which contact interactions play a much more important role that in the original Weinberg counting (though this is in general only true at the lowest orders). There are still a few minor details that are not completely understood in what regard the renormalization of repulsive singular interactions, which explains the small mismatches in power counting in the three independent works that have analyzed it.[10–15]

We show the EFT predictions for the S-wave phase shifts in Fig. 1. A few details of the calculation are explained in the caption, while further details plus the P- and D-wave phase shifts can be found in Refs. 10,11. In general the quality of the description is relatively good in comparison with that of the Weinberg counting[16,17] at the same order, which serves as a cross-check that we are going in the right direction. In the future, concerning the two-body sector we have to go to higher orders to achieve a high-quality description of the phase shifts. Beyond two-body, nuclear EFT has to be extended to reactions on the deuteron (form factors, neutron capture, photo- and electrodisintegration), to the few-body sector and eventually to nuclear matter.

---

[a] Actually this is an overly simplified version of the true renormalization process. Reality is much more messy.

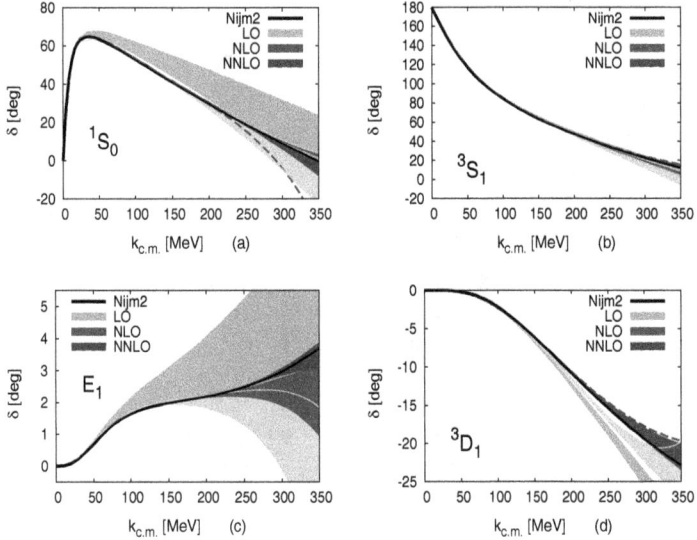

Fig. 1. The S-wave phase shifts in nuclear EFT, with δ the phase shift and $k_{c.m.}$ the center-of-mass momentum. The bands reflect the cut-off uncertainty of the results in the range $r_c = 0.6 - 0.9\,\mathrm{fm}$. We compare with the NNLO phase shifts in the Weinberg counting (the light blue band).[16,17]

References

1. G. A. Miller, B. M. K. Nefkens and I. Slaus, *Phys. Rept.* **194**, 1 (1990).
2. S. Beane, P. Bedaque, K. Orginos and M. Savage, *Phys.Rev.Lett.* **97**, p. 012001 (2006).
3. N. Ishii, S. Aoki and T. Hatsuda, *Phys.Rev.Lett.* **99**, p. 022001 (2007).
4. V. Bernard, N. Kaiser and U.-G. Meissner, *Int.J.Mod.Phys.* **E4**, 193 (1995).
5. S. Weinberg, *Phys. Lett.* **B251**, 288 (1990).
6. S. Weinberg, *Nucl. Phys.* **B363**, 3 (1991).
7. A. Nogga, R. G. E. Timmermans and U. van Kolck, *Phys. Rev.* **C72**, p. 054006 (2005).
8. M. Pavon Valderrama and E. Ruiz Arriola, *Phys. Rev.* **C74**, p. 054001 (2006).
9. M. Pavon Valderrama and E. Ruiz Arriola, *Phys. Rev.* **C74**, p. 064004 (2006).
10. M. Pavon Valderrama, *Phys.Rev.* **C83**, p. 024003 (2011).
11. M. Pavon Valderrama, *Phys.Rev.* **C84**, p. 064002 (2011).
12. M. C. Birse, *Phys. Rev.* **C74**, p. 014003 (2006).
13. B. Long and C. Yang, *Phys.Rev.* **C84**, p. 057001 (2011).
14. B. Long and C. Yang, *Phys.Rev.* **C85**, p. 034002 (2012).
15. B. Long and C. Yang, *Phys.Rev.* **C86**, p. 024001 (2012).
16. E. Epelbaum, W. Glöckle and U.-G. Meißner, *Eur. Phys. J.* **A19**, 125 (2004).
17. E. Epelbaum, W. Glöckle and U.-G. Meißner, *Eur. Phys. J.* **A19**, 401 (2004).

Present status of light scalars

J. R. Peláez

Departamento de Física Teórica II. Facultad de CC. Físicas.
Universidad Complutense. 28040 Madrid. Spain
E-mail: jrpelaez@fis.ucm.es

I will briefly review in this talk the most significant developments in light scalars spectroscopy that have occurred since the last edition of this Conference, although I will comment on the updates of the mass and width of different light scalar mesons in the latest edition of the Particle Data Tables, I will be mostly focusing on the major revision of the $f_0(500)$ meson, also known as σ meson. I will explain how this major update has been driven both by new data and rigorous dispersive analyses.

Keywords: Light scalar mesons.

1. Introduction

Light scalar mesons play a prominent role in the nucleon-nucleon attractive interaction, the QCD spontaneous chiral symmetry breaking as well as in the search for glueballs. Despite this relevance, light scalars have suffered a longstanding controversy concerning their properties, spectroscopic classification and even their very existence as it has been nicely summarized in the "Note on light scalars below 2 GeV" in the Review of Particle Properties (RPP).[1] In this talk I briefly review how the combination of new data with rigorous and model independent approaches has finally provided very convincing evidence of the existence and properties of these states. Very slowly, these developments are cautiously being reflected in the RPP. Actually in the latest RPP edition the σ meson has suffered a major revision and the f0(980) has been significantly updated.

Here I will report on the progress made after the previous 2010 International Symposium on Chiral Symmetry in Hadrons and Nuclei, concentrating on mass and width determinations of scalars below 1 GeV. The debate over their composition and classification in multiplets together with other heavier scalars, lie beyond the scope of this talk. I will follow two approaches: a conservative one, based on the RPP most recent changes, and my personal opinion, less conservative but closer to that of the "scalar community", which for long has been well aware of light scalar meson status now acknowledged by the RPP. I will emphasize that the RPP updates have been driven by new data, but most importantly by the rigour and consistency of dispersive analyses. Since similar results exist for other light scalars, I expect further major updates in the near future for other resonances, particularly the $K(800)$.

2. The σ or $f_0(500)$ Meson. A Major Change in the RPP

It is already 60 years ago that a light scalar-isoscalar field was postulated[2] in order to explain the inter-nucleon attraction. To describe chiral symmetry in pion-pion interactions, this field was soon included within the Linear Sigma Model,[3] from which it gets its usual name: the σ meson, its modern name being $f_0(500)$. The linear sigma model is a simple and linear realization of an spontaneous chiral symmetry breaking, where all fields but the σ become Goldstone bosons, i.e. pions. Generically, the σ, having the vacuum quantum numbers, plays a relevant role for the understanding of the QCD spontaneous chiral symmetry breaking.

The latest RPP revision of the σ meson can be considered a major improvement in view of its history: until 1974 the RPP listed the σ as "not-well established", from 1976 it then disappeared for 20 years and came back as the $f_0(600)$ in 1996. The reason for this coming and going is that the nucleon-nucleon interaction is rather insensitive to the details of the exchanged particles, and even less so if they are as wide as the σ. Therefore, light scalars were mostly studied in meson-meson scattering, where they can be produced in the s-channel. Unfortunately, $\pi\pi$ scattering is extracted from $\pi N \to \pi\pi N$ through a complicated analysis full of systematic uncertainties, and the experimental results[4] were often in conflict. Take a look, for instance, at Fig.1, which shows the scalar-isoscalar $\pi\pi$ scattering phase, and check the large differences between data sets,[4] even within the same collaboration. The $f_0(600)$ case gathered enough support before 2002 when it was declared "well established" by the RPP. This came from some theoretical works but mostly from the results of heavy meson decay experiments. Surprisingly for a "well established" state, it was quoted with a huge uncertainties: from 400 to 1200 MeV fro the mass, and from 500 to 1000 MeV, for the width. These huge ranges and the $f_0(600)$ name were kept until the last 2010 RPP edition.

Back to Fig.1 note that, on the one hand, the data below 400 MeV coming from $K \to \pi\pi\ell\nu$ decays,[5,6] have almost no systematic uncertainty compared to those from $\pi N \to \pi\pi N$. Particularly important are the NA48/2 precise data from 2010,[6] since being consistent with them is a very important criterion for the RPP choice of results in their 2012 estimate. On the other hand, there is nothing like a Breit–Wigner shape around 500-600 MeV. In fact, the σ cannot be described as a Breit–Wigner resonance (nor the $K_0^*(800)$), but one has to use the mathematically rigorous definition by means of its associated pole in the complex plane, whose position s_R is related to the resonance mass and width as $\sqrt{s_R} \simeq M_R - i\Gamma_R/2$. For this reason the RPP provides a "t-matrix" pole, although, unfortunately, it also provides a Breit–Wigner pole. Obviously, the latter only leads to confusion, and thus I will only comment "t-matrix" poles. Hence, in Fig. 2 displays the position of the σ poles in the RPP and the light gray area stands for the huge RPP uncertainty estimate until 2010.

It is important to remark that to determine poles deep in the complex plane, it is not enough to have a good data description, a correct analytic continuation is really needed. Unfortunately this has not been the case of many analyses, leading to unreliable or plain wrong pole determinations. As a matter of fact, most of the disagreement seen in Fig. 2 can be attributed to unreliable analytic extrapolations, so that even the same experiment can provide surprisingly different poles. For instance, the poles at $400 - i500$ MeV, $1100 - i300$ MeV and $1100 - i137$ MeV (below the legend), both come from Ref. 7. The lesson to learn is that only poles extracted from rigorous analytic or dispersive approaches provide reliable σ pole determinations, which in Fig. 2 are plotted in colors other than red. By looking only at those poles it is clear why the existence of a σ pole around 500 MeV was rather well known for many years within the scalar meson community. Dispersive approaches may differ by few tens of MeV, but definitely not by several hundreds. Note that poles obtained from heavy meson decays (with

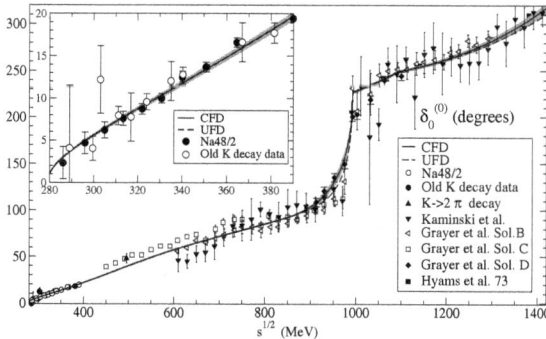

Fig. 1. Scalar-isoscalar $\pi\pi$ scattering phase $\delta_0^{(0)}$[4–6] versus the UFD and CFD parametrization.[13]

no updates in the 2012 RPP[1]) yield a somewhat higher mass than dispersive approaches, between 500 and 550 MeV. Let me also remark that the analysis of these decays has been frequently performed with models much less rigorous than dispersive analyses.

A correct analytic continuation is obtained from dispersion relations. These are just a consequence of causality and relate the amplitude at any value with its integral over the real axis, i.e. the data. Thanks to the integral representation the results are independent of the model or functional form parameterizing the data. Dispersion relations can be used to: (i) check the data consistency at a given energy versus data in other regions, (ii) constrain the fits to data, (iii) obtain the value of the amplitude at energies where data are not available, (iv) implement the analytic continuation in order to look for poles. Particularly useful for spectroscopy are partial wave dispersion relations, since their poles are directly associated to resonances with their same quantum numbers. Unfortunately, due to crossing symmetry, partial waves have additional singularities from the unphysical s region, like the "left cut" contribution or even a circular one in case of unequal masses. These may be numerically relevant for precise studies of the σ and the $K_0^*(800)$, which are relatively close to threshold and the left cut. Dealing rigorously with the left cut requires an infinite set of coupled integral equations, known as Roy equations[8] for $\pi\pi$ scattering, which have shown to be very powerful, particularly over the last decade.[9–15] The reason is that in the 70's, their accuracy was limited by the quality of threshold data, but this caveat can be circumvented either by means of Chiral Perturbation Theory (ChPT) at low energy, as in Ref. 9, or, if one wants to avoid ChPT as in Ref. 13, by using the recent and precise NA48/2 data.[6] The former approach yielded a precise σ pole in,[12] where it was also shown that Roy eqs. provide a consistent analytic continuation down to the σ pole. The latter approach is merely a dispersive data analysis, was carried out by our group[13,14] using another set of Roy-like Equations with only one subtraction (less energy suppression in the integrals), called GKPY Equations.

Thus, the 2012 RPP has finally made a major revision of the σ mass range, reduced by a factor of more than five, down to 400 to 550 MeV, and by almost a factor of two for the width, now estimated between 400 and 700 MeV. This change has been triggered by the consistency of dispersive results, together with the new NA48/2 data close to $\pi\pi$ threshold.[6] The new RPP "estimate", shown in Fig. 2 as the smaller and darker rectangle, takes into account, not only the most recent dispersive analyses, but other

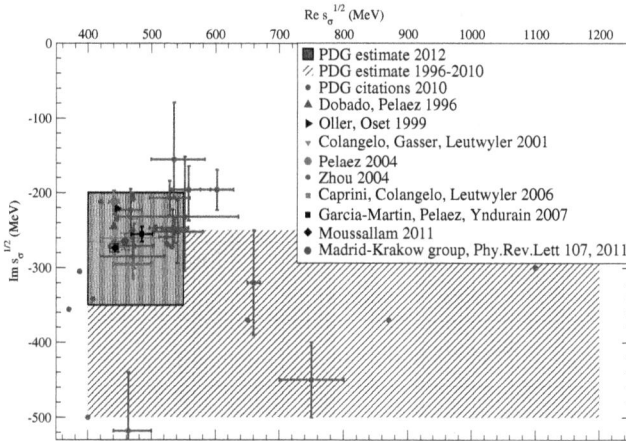

Fig. 2. (Color online) $f_0(500)$ poles in the RPP.[1] Non-red poles, obtained from dispersive or analytic approaches,[9,12,14,15,18] fall within the 2012 estimate, which is a major revision with respect to the 2010 RPP estimate.

results from models which are required to be at least consistent with the accurate $K \to \pi\pi\ell\nu$ data,[6,16] as well as values from other processes like heavy meson decays, which, as commented above, use models and yield somewhat larger masses than dispersive approaches. Note that even the name of the particle has been changed to $f_0(500)$. The RPP also provides Breit–Wigner parameters but I have argued above why I think they should be avoided. Definitely, this should be considered a major revision constitutes which was long awaited, and improves considerably the previous situation. Nevertheless, in my opinion, these RPP criteria are still too conservative, and for the σ I would rely on pole extractions based on rigorous dispersive techniques only. In fact, even the RPP "Note on light scalars" offers the possibility to "take the more radical point of view and just average the most advanced dispersive analyses" (here[9,12,14,15]), yielding: $\sqrt{s_\sigma} = (446 \pm 6) - (276 \pm 5)$ MeV.

As an illustration of the dispersive techniques, let me sketch the methods of our group.[13,14] We start from a set of "Unconstrained Fits to Data" (UFD), which we showed not to be too inconsistent with forward dispersion relations. Next, we slightly modify the fit to satisfy dispersion relations without spoiling the description of the experimental data. This procedure leads to a set of "Constrained Fits to Data" (CFD). Both the CFD and UFD for the scalar-isoscalar $\pi\pi$ scattering phase shift are shown on Fig.1. The only noticeable differences between the UFD and CFD appear in or above the 1 GeV region, but both sets describe data very well. Let us remark that, given the same input, in the resonance region above 500 MeV the once-subtracted GKPY equations are more precise, whereas Roy equations are more accurate below that energy. In summary, the CFD describes the data and is consistent with a whole set of dispersion relations, unitarity and symmetry constraints, etc. We then use this CFD inside the dispersion relation to obtain the analytical continuation of the partial wave into the complex plane, where the following poles are found:[14]

$$\sqrt{s_\sigma} = (457^{+14}_{-13}) - i(279^{+11}_{-7}) \text{ MeV} \quad \text{(from GKPY eqs.)}$$

and $(445 \pm 25) - i(278^{+22}_{-18})$ MeV (from Roy eqs.).

These results from our group are two of the five new entries in the 2012 RPP edition. The other ones are two results from an "analytic K-matrix model" in Ref. 17: $(452 \pm 13) - i(259 \pm 16)$ MeV and $(448 \pm 43) - i(266 \pm 43)$ MeV, which differ on what data sets and different variants of the K-matrix model are averaged, together with $(442^{+5}_{-8}) - i(274^{+6}_{-5})$ MeV from Ref. 15. The latter is based on Roy equations using as input for the other waves and higher energies the Roy equations output of Ref. 9; hence, it is very consistent with the older result in Ref. 12: $\sqrt{s_\sigma} = (441^{+16}_{-8}) - i(272^{+9}_{-12.5})$ MeV, which used ChPT as input. These last two are also consistent with the older result in Ref. 10: $(452 \pm 13) - i(259 \pm 16)$ MeV. Note that these last three results, based on Roy equations, together with our two results in the paragraph above, also based on dispersive approaches, are the ones that the RPP considers as the "most advances dispersive analyses".

3. The Other Light Scalars

The existence and parameters of the $f_0(980)$ and $a_0(980)$ have been much less controversial, since these resonances are narrow and clearly seen in many processes. For example, in Fig.1, although it is slightly distorted by the nearby $\bar{K}K$ threshold, a Breit–Wigner-like shape over a background phase of about 100° around 980 MeV. This is nothing but the $f_0(980)$.

There were no changes for the $a_0(980)$ in the latest RPP. However, after almost two decades keeping the same estimate, the 2012 RPP edition has updated the $f_0(980)$ mass to 990 ± 20 MeV from the old value of 980 ± 10 MeV. In the "Note on light scalars", it is explained that the mass shift of 10 MeV upwards, and the doubling of the uncertainty, was made to include our dispersive analysis.[13,14] Actually, we obtain an $f_0(980)$ second Riemann sheet pole at: $\sqrt{s_{f_0(980)}} = (996 \pm 7) - i(25^{+10}_{-6})$ MeV when using GKPY

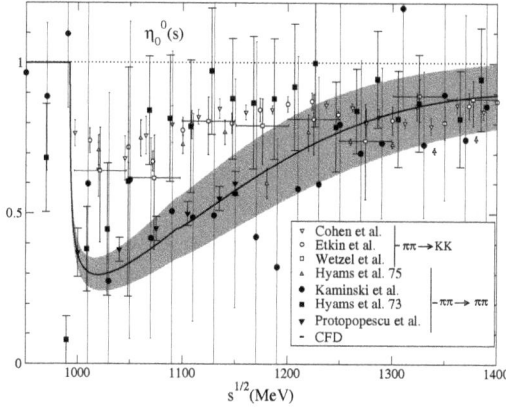

Fig. 3. Favored "Dip" solution (CFD) for the $\pi\pi$ scattering inelasticity around the $f_0(980)$ resonance region. Original figures and references in Ref. 13.

equations and $\sqrt{s_{f_0(980)}} = (1003^{+5}_{-27}) - i(21^{+10}_{-8})$ MeV from Roy equations. The relevance of our result is that it has settled a longstanding controversy between the "dip" and "no-dip" sets of data for the elasticity parameter in $\pi\pi$ scattering, shown in Fig. 3. In particular, we showed that the dip scenario is consistent with the GKPY equations as seen in Fig. 3, whereas the non-dip scenario was strongly disfavored. Later on this was confirmed in Ref. 15 using Roy equations, also obtaining a pole at: $(996^{+4}_{-14}) - i(24^{+11}_{-3})$ MeV. In fact, these three dispersive values together with $(981 \pm 43) - i(18 \pm 11)$ MeV as obtained in Ref. 17 using the "analytic K-matrix" approach, are the only new $f_0(980)$ entries in the 2012 RPP edition.

The RPP list the $K_0^*(800)$ meson or κ, under the "needs confirmation" label. This occurs despite being rather similar to the σ. In fact either within fairly reasonable models or rigorous Roy-like dispersive analyses,[19] the $K_0^*(800)$ appears as a wide pole (not a Breit Wigner) around 650 to 770 MeV, with a 550 MeV width or larger. Furthermore, as with the $f_0(500)$, a pole is consistently seen in heavy meson decays. In my opinion then, it deserves the same treatment as the σ and should be considered as another "well established" light scalar meson resonance.

Unfortunately, the new entries in the 2012 RPP edition come from Breit–Wigner parameters of two studies of J/Ψ decays at BES2.[20] Nevertheless this Collaboration[20] also provides a t-matrix pole position $764 \pm 63^{+71}_{-54} - i(306 \pm 149^{+143}_{-85})$ MeV. The value of this pole is very consistent with the one from Roy-like dispersion relations $(658 \pm 13) - i(278.5 \pm 12)$ MeV.[19] Still, the 2012 RPP keeps the previous estimates: 685 ± 29 MeV for the mass and 547 ± 24 for the width.

Acknowledgments

Let me thank the organizers for creating such a comfortable and inspiring atmosphere for discussing Physics, in particular I would very much like to thank Lisheng Geng and Binsong Zou for their kind invitation, their patience, kind help as well as the hospitality of the Beihang University. Research partially funded by the Spanish Research contract FPA2011-27853-C02-02 and the EU FP7 HadronPhysics3 project.

References

1. J. Beringer et al. (Particle Data Group), Phys. Rev. D86, 010001 (2012).
2. M. H. Johnson and E. Teller, Phys. Rev. **98**, 783 (1955).
3. M. Gell-Mann and M.Levy, Nuovo Cim. **16**, 705 (1960).
4. Protopopescu, S. D., et al., *Phys Rev.* **D7**, 1279, (1973). Grayer, G., *et al.*, *Nucl. Phys.* **B75**, 189, (1974). Losty, M. J., *et al. Nucl. Phys.*, **B69**, 185 (1974). Hyams, B., *et al.*, *Nucl. Phys.* **B100**, 205, (1975). P. Estabrooks and A. D. Martin, Nucl. Phys. B **79**, 301 (1974). R. Kaminski, L. Lesniak and K. Rybicki, Z. Phys. C **74**, 79 (1997).
5. L. Rosselet, *et al.*, Phys. Rev. **D15**, 574 (1977). S. Pislak *et al.*, Phys. Rev. Lett. **87** (2001) 221801.
6. J. R. Batley *et al.* [NA48-2 Collaboration], Eur. Phys. J. **C70**, 635-657 (2010).
7. C. Amsler *et al.* Phys. Lett. B **355**, 425 (1995) and Phys. Lett. B **342**, 433 (1995).
8. S. M. Roy, Phys. Lett. B **36**, 353 (1971).
9. G. Colangelo, J. Gasser and H. Leutwyler, Nucl. Phys. B **603**, 125 (2001).
10. B. Ananthanarayan, G. Colangelo, J. Gasser and H. Leutwyler, Phys. Rept. **353**, 207 (2001).
11. R. Kaminski, L. Lesniak and B. Loiseau, Phys. Lett. B **551**, 241 (2003).
12. I. Caprini, G. Colangelo and H. Leutwyler, Phys. Rev. Lett. **96**, 132001 (2006).
13. R. Garcia-Martin, *et al.*, Phys. Rev. D **83**, 074004 (2011).
14. R. Garcia-Martin, *et al.*, Phys. Rev. Lett. **107**, 072001 (2011).
15. B. Moussallam, Eur. Phys. J. C **71**, 1814 (2011).
16. S. Pislak *et al.* [BNL-E865 Collaboration], Phys. Rev. Lett. **87** (2001) 221801.
17. G. Mennessier, S. Narison and X. G. Wang, Phys. Lett. B **688**, 59 (2010).
18. A. Dobado and J. R. Pelaez, Phys. Rev. D **56**, 3057 (1997). J. A. Oller and E. Oset, Phys. Rev. D **60** (1999) 074023. J. A. Oller, E. Oset and J. R. Pelaez, Phys. Rev. D **59** (1999) 074001; [Erratum-ibid. D **60** (1999) 099906]; [Erratum-ibid. D **75** (2007) 099903]. J. R. Pelaez, Mod. Phys. Lett. A **19**, 2879 (2004) Z. Y. Zhou *et al.*, JHEP **0502**, 043 (2005). R. Garcia-Martin and J. R. Pelaez, F. J. Yndurain, Phys. Rev. D **76**, 074034 (2007)
19. S. Descotes-Genon and B. Moussallam, Eur. Phys. J. C **48**, 553 (2006).
20. M. Ablikim, *et al.*, Phys. Lett. B **693**, 88 (2010) and Phys. Lett. B **698**, 183 (2011).

Hyperon-nucleon interaction and baryonic contact terms in SU(3) chiral effective field theory

S. Petschauer

Physik Department, Technische Universität München,
D-85747 Garching, Germany
E-mail: stefan.petschauer@ph.tum.de

In this paper we summarize results for baryonic contact terms derived within SU(3) chiral effective field theory. The four-baryon contact terms, necessary for the description of the hyperon-nucleon interaction, include SU(3) symmetric and explicit chiral symmetry breaking terms. They also include four-baryon contact terms involving pseudoscalar mesons, which become important for three-body forces. Furthermore we derive the leading order six-baryon contact terms in the non-relativistic limit and study their contribution to the ΛNN three-body contact interaction. These results could play an important role in studies of hypernuclei or hyperons in nuclear matter.

Keywords: SU(3) chiral effective field theory, two- and three-baryon forces.

1. Introduction

The nuclear forces are very well described within the framework of SU(2) chiral effective field theory.[1–3] Therefore it is natural to extend this scheme to the strangeness sector and use SU(3) chiral effective field theory to describe the interaction between baryons, as has been done in a recent calculation of the hyperon-nucleon interaction at next-to-leading order.[4] There the unresolved short-distance dynamics is encoded in four-baryon contact terms with *a priori* unknown low-energy constants. These contact terms are constructed according to the symmetries of QCD.

It is not only interesting to understand baryon-baryon scattering itself, but these interactions are also input for studies of hypernuclei and hyperons in nuclear matter, such as exotic neutron star matter. Especially for the few- and many-body calculations, not only two-baryon interactions, but also three-baryon interactions will play an important role.[5–7] To support the recently observed two-solar-mass neutron stars,[8,9] a very stiff equation of state and therefore a repulsive hyperon-nucleon force is needed. In order to

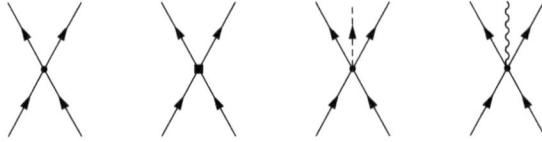

Fig. 1. Examples for four-baryon contact vertices at LO and NLO. The dashed and wavy lines denote pseudoscalar mesons and photons, respectively.

achieve enough repulsion, it might be necessary to include hyperon-nucleon-nucleon three-body forces as well. As a first step the leading three-baryon contact forces have been derived and classified within SU(3) chiral effective field theory.

2. Four-baryon Contact Terms

Considering the baryon-baryon interaction, at leading order one has non-derivative four-baryon contact terms and at next-to-leading order four-baryon contact terms with two derivatives (Fig. 1), both SU(3) symmetric. Additionally, at next-to-leading order pure baryon-baryon contact terms proportional to quark masses arise and lead to explicit SU(3) symmetry breaking. Furthermore, at next-to-leading order occur four-baryon contact terms involving one or more pseudoscalar mesons and/or external electroweak fields, cf. Fig. 1. These come into play in the description of chiral many-body forces and currents relevant for few-baryon systems.

By employing the external fields method,[10,11] we have constructed a complete set of terms for the covariant chiral baryon-baryon contact Lagrangian in flavor SU(3) up to order $\mathcal{O}(q^2)$,[12] which provides the above mentioned contact terms. The constructed terms are invariant under charge conjugation, parity transformation, time reversal, Hermitian conjugation and local chiral transformations and include Goldstone bosons as well as external fields. In the case of pure baryon-baryon interaction one obtains a minimal set of 40 terms in the chiral contact Lagrangian up to $\mathcal{O}(q^2)$. After a non-relativistic reduction and a decomposition into partial waves, 28 of these contact terms lead to SU(3) symmetric contributions to the potentials for the channels $^1S_0$, $^3S_1$, $^1P_1$, $^3P_0$, $^3P_1$, $^3P_2$, $^3D_1 \leftrightarrow {}^3S_1$ and $^1P_1 \leftrightarrow {}^3P_1$. Only one specific term leads to an antisymmetric spin-orbit interaction and therefore a spin singlet-triplet mixing. Such transitions are not possible in the NN interaction without SU(2) symmetry breaking. The remaining 12 low-energy constants contribute to the $^1S_0$ and $^3S_1$ partial waves and lead to SU(3) symmetry breaking contributions linear in the quark masses.

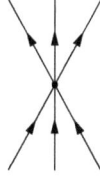

Fig. 2.   Feynman diagram for the leading three-baryon contact force.

3. Six-baryon Contact Terms

To describe the effect of three-body forces the full set of the leading three-baryon interactions with a minimal number of low-energy constants, consistent with the symmetries of QCD, is needed. In contrast to phenomenological approaches,[5,6] we want to construct these three-baryon forces within the framework of SU(3) chiral effective field theory (following the construction of the chiral nuclear forces[1]). We start this construction by deriving the short-range contact contribution in SU(3) chiral effective field theory in Fig. 2.

The construction of the chiral Lagrangian works analogously to the construction of the four-baryon contact terms. The (approximate) symmetries of QCD have to be fulfilled namely charge conjugation, parity transformation, time reversal, Hermitian conjugation and chiral symmetry. For the construction of the potential it is also important to include the Pauli exclusion principle, since one starts with an overcomplete set of terms in the Lagrangian and wants to obtain the minimal number of low-energy constants. This can be achieved by an antisymmetrization in both the initial and the final state. The Pauli principle is not as restrictive as for the three-nucleon contact interaction, where one ends up with only one low-energy constant, but it is still relevant.

After constructing the pertinent terms in the chiral Lagrangian and performing a non-relativistic reduction, the leading potentials can be expressed through the following operators in spin space

$$\mathbb{1}, \quad \vec{\sigma}_1 \cdot \vec{\sigma}_2, \quad \vec{\sigma}_1 \cdot \vec{\sigma}_3, \quad \vec{\sigma}_2 \cdot \vec{\sigma}_3, \quad i\vec{\sigma}_1 \times \vec{\sigma}_2 \cdot \vec{\sigma}_3 .$$

For each three-baryon channel the prefactors of these spin operators are a combination of SU(3) coefficients and low-energy constants.

As a result we have obtained that the symmetry constraints lead to 18 low-energy constants for the three-baryon contact force. Table 1 gives the number of additional constants that are introduced with increasing strangeness of the three-baryon system. As an explicit example we give the

Table 1. Results for the low-energy constants of the leading three-baryon contact terms.

strangeness	transitions between	# constants
0	NNN	1
−1	$\Lambda NN, \Sigma NN$	+7
−2	$\Lambda\Lambda N, \Lambda\Sigma N, \Sigma\Sigma N, \Xi NN$	+9
−3	$\Lambda\Lambda\Sigma, \Lambda\Sigma\Sigma, \Sigma\Sigma\Sigma, \Xi\Lambda N, \Xi\Sigma N$	+1
−4	$\Xi\Xi N, \Xi\Lambda\Lambda, \Xi\Lambda\Sigma, \Xi\Sigma\Sigma$	+0
−5	$\Xi\Xi\Lambda, \Xi\Xi\Sigma$	+0
−6	$\Xi\Xi\Xi$	+0
		18

form of the potentials for the ΛNN interaction with isospin 0 and 1:

$$V_{\Lambda NN \to \Lambda NN}^{I=0} = e_2(\mathbb{1} + \tfrac{1}{3}\vec{\sigma}_2 \cdot \vec{\sigma}_3) + e_3\left(\vec{\sigma}_1 \cdot \vec{\sigma}_2 + \vec{\sigma}_1 \cdot \vec{\sigma}_3\right),$$
$$V_{\Lambda NN \to \Lambda NN}^{I=1} = e_4(\mathbb{1} - \vec{\sigma}_2 \cdot \vec{\sigma}_3).$$

Note that the low-energy constants for these two isospin channels are independent. The constant e_1 (proportional to c_E, which is present in the purely nucleonic sector[1]) does not appear in the ΛNN interaction. Therefore the ΛNN contact interaction can not be constrained by the three-nucleon sector.

Acknowledgments

This work has been supported in part by DFG and NSFC (CRC110) and the "TUM Graduate School". I thank J. Haidenbauer, N. Kaiser, U.-G. Meißner, A. Nogga and W. Weise for fruitful collaboration.

References

1. E. Epelbaum et al., *Rev.Mod.Phys.* **81**, 1773 (2009).
2. R. Machleidt and D. Entem, *Phys.Rept.* **503**, 1 (2011).
3. J. W. Holt, N. Kaiser and W. Weise, *Prog.Part.Nucl.Phys.* **73**, 35 (2013).
4. J. Haidenbauer et al., *Nucl.Phys.* **A915**, 24 (2013).
5. R. Bhaduri, B. Loiseau and Y. Nogami, *Annals Phys.* **44**, 57 (1967).
6. A. Gal, J. Soper and R. Dalitz, *Annals Phys.* **63**, 53 (1971).
7. D. Lonardoni, S. Gandolfi and F. Pederiva, *Phys.Rev.* **C87**, 041303 (2013).
8. P. B. Demorest et al., *Nature* **467**, 1081 (2010).
9. J. Antoniadis et al., *Science* **340**, 6131 (2013).
10. J. Gasser and H. Leutwyler, *Annals Phys.* **158**, 142 (1984).
11. J. Gasser and H. Leutwyler, *Nucl.Phys.* **B250**, 465 (1985).
12. S. Petschauer and N. Kaiser, *Nucl.Phys.* **A916**, 1 (2013).

Octet baryon masses and sigma terms in covariant baryon chiral perturbation theory

Xiu-Lei Ren* and Li-Sheng Geng†

School of Physics and Nuclear Energy Engineering and International Research Center for Nuclei and Particles in the Cosmos, Beihang University, Beijing, 100191, China
* E-mail: xiuleiren@phys.buaa.edu.cn
† E-mail: lisheng.geng@buaa.edu.cn

Jie Meng

School of Physics and Nuclear Energy Engineering and International Research Center for Nuclei and Particles in the Cosmos, Beihang University, Beijing, 100191, China
State Key Laboratory of Nuclear Physics and Technology, School of Physics, Peking University, Beijing, 100871, China
Department of Physics, University of Stellenbosch, Stellenbosch, 7602, South Africa
E-mail: mengj@pku.edu.cn

We report an analysis of the octet baryon masses using the covariant baryon chiral perturbation theory up to next-to-next-to-next-to-leading order with and without the virtual decuplet contributions. Particular attention is paid to the finite-volume corrections and the finite lattice spacing effects on the baryon masses. A reasonable description of all the publicly available $n_f = 2 + 1$ lattice QCD data is achieved. Utilyzing the Feynman-Hellmann theorem, we determine the nucleon sigma terms as $\sigma_{\pi N} = 55(1)(4)$ MeV and $\sigma_{sN} = 27(27)(4)$ MeV.

Keywords: Chiral lagrangians; lattice QCD calculations; Baryon resonances.

PACS numbers: 12.39.Fe, 12.38.Gc, 14.20.Gk

1. Introduction

Recently, the lowest-lying octet baryon masses have been studied on the lattice with $n_f = 2 + 1$ configurations.[1-9] Because the limitation of the computational resources, most lattice quantum chromodynamics (LQCD) simulations still have to employ larger than physical light-quark masses, finite lattice volume and finite lattice spacing. Chiral perturbation theory (ChPT),[10] as an effective field theory of low-energy QCD, plays an important role in performing the multiple extrapolations needed to extrapolate LQCD results (chiral extrapolations,[11–14] finite-volume corrections (FVCs),[15,16] and continuum extrapolations[17,18]) to the physical world.

In this work we report on the first systematic study of the ground-state octet baryon masses in the covariant baryon chiral perturbation theory (BChPT) with the extended-on-mass-shell (EOMS) scheme up to next-to-next-to-next-to-leading order (N³LO). The virtual decuplet contributions to the octet baryon masses and finite lattice volume and lattice spacing effects on the lattice data are studied. Finally, the octet baryon sigma terms are predicted using the Feynman-Hellmann theorem.

2. Theoretical Framework

Up to N³LO, the octet baryon masses with the virtual decuplet contributions can be written as

$$m_B = m_0 + m_B^{(2)} + m_B^{(3)} + m_B^{(4)} + m_B^{(D)},\tag{1}$$

where m_0 is the chiral limit octet baryon mass, $m_B^{(2)}$, $m_B^{(3)}$, and $m_B^{(4)}$ correspond to the $\mathcal{O}(p^2)$, $\mathcal{O}(p^3)$, and $\mathcal{O}(p^4)$ contributions from the octet-only EOMS BChPT, respectively. The last term $m_B^{(D)}$ denotes the contributions of the virtual decuplet resonances up to N³LO. Their explicit expressions and the corresponding FVCs can be found in Refs.[19,20]

In order to perform the continuum extrapolation of the LQCD simulations, one can first write down the Symanzik's effective filed theory.[21,22] In Ref.,[23] we constructed the corresponding chiral Lagrangians up to $\mathcal{O}(a^2)$ to study the finite lattice spacing effects on the octet baryon masses, which can be written as

$$m_B^{(a)} = m_B^{\mathcal{O}(a)} + m_B^{\mathcal{O}(am_q)} + m_B^{\mathcal{O}(a^2)}.\tag{2}$$

Here we want to mention that there are 19 unknown LECs (m_0, b_0, b_D, b_F, $b_{1,\cdots,8}$, $d_{1,\cdots,5,7,8}$) needed to be fixed in the EOMS BChPT at $\mathcal{O}(p^4)$. Furthermore, including the finite lattice spacing effects (Eq. (2)), one has to introduce 4 more combinations of the unknown LECs.[23]

3. Results and Discussions

The details of the studies can be found in Refs. 19,20,23,24. Here we only briefly summarize the main results.

In order to determine all the LECs and test the consistency of the current LQCD simulations, we perform a simultaneous fit to all the publicly available $n_f = 2 + 1$ LQCD data from the PACS-CS,[3] LHPC,[5] QCDSF-UKQCD,[8] HSC,[6] and NPLQCD[9] Collaborations. To ensure that the N³LO

BChPT stays in its applicability range, fitted LQCD data are limited to those satisfying $M_\pi^2 < 0.25$ GeV^2 and $M_\phi L > 4$.

In Refs. 19,20, we found that the octet-only EOMS BChPT shows a good description of the LQCD and experimental data with order-by-order improvement. Up to N^{3}LO, the χ^2/d.o.f. is about 1.0, which indicates that the lattice simulations from these five collaborations are consistent with each other *, although their setups are very different. In addition, we showed that the explicit inclusion of the virtual decuplet baryons does not change the description of the LQCD data in any significant way, at least at $\mathcal{O}(p^4)$. This implies that using only the octet baryon mass data, one can not disentangle the virtual decuplet contributions from those of the virtual octet baryons and tree-level diagrams. On the other hand, we notice that the explicit inclusion of the virtual decuplet baryons does seem to improve slightly the description of the FVCs, especially for the LQCD data with small $M_\phi L$. Therefore, the virtual decuplet contributions to the octet baryon masses are not taken into account in our following studies.

To study discretization effects on the ground-state octet baryon masses, we constructed the relevant chiral Lagrangians up to $\mathcal{O}(a^2)$ in Ref. 23. By analyzing the latest $n_f = 2+1$ $\mathcal{O}(a)$-improved LQCD data of the PACS-CS, QCDSF-UKQCD, HSC and NPLQCD Collaborations, we found that the finite lattice spacing effects are at the order of $1 - 2\%$ for lattice spacings up to 0.15 fm and the pion mass up to 500 MeV, which is in agreement with other LQCD studies.

Finally, the octet baryon sigma terms are predicted using the Feynman-Hellmann theorem. In order to obtain an accurate determination of sigma terms, a careful examination of the LQCD data is essential, since not all of them are of the same quality though they are largely consistent with each other. In Ref. 24, we only employed the PACS-CS, LHPC and QCDSF-UKQCD data. We also took into account the scale setting effects of the LQCD simulations and studied systematic uncertainties from truncating chiral expansions. Furthermore, strong-interaction isospin breaking effects to the baryon masses were for the first time employed to better constrain the relevant LECs up to N^{3}LO. We predict the nucleon sigma terms as $\sigma_{\pi N} = 55(1)(4)$ MeV and $\sigma_{sN} = 27(27)(4)$ MeV, which are consistent with recent LQCD and BChPT studies.

---

*This does not seem to be the case for the LQCD simulations of the ground-state decuplet baryon masses.[25]

4. Conclusions

We have studied the lowest-lying octet baryon masses in the EOMS BChPT up to N^3LO. The unknown low-energy constants are determined by a simultaneous fit to the latest $n_f = 2 + 1$ LQCD simulations, and it is shown that the LQCD results are consistent with each other, though their setups are quite different. The contributions of virtual decuplet resonances are explicitly included and we find that their effects on the octet baryon masses are small, especially for the chiral extrapolations.

We have studied finite-volume corrections and finite lattice spacing effects on the LQCD baryon masses as well. We find that their effects are of similar size but finite volume corrections are more important to better constrain the LECs and to reduce the $\chi^2/$d.o.f..

Using the Feynman-Hellmann theorem, we have performed an accurate determination of the nucleon sigma terms, focusing on the uncertainties from the lattice scale setting method and chiral expansions. Our predictions are $\sigma_{\pi N} = 55(1)(4)$ MeV and $\sigma_{sN} = 27(27)(4)$ MeV, which are consistent with most of the recent LQCD and BChPT studies. However, further LQCD simulations are needed to reduce the uncertainty of the nucleon strangeness-sigma term.

Acknowledgments

X.-L. R. acknowledges the Innovation Foundation of Beihang University for Ph.D. Graduates. This work was partly supported by the National Natural Science Foundation of China under Grants No. 11005007, No. 11375024, and No. 11175002, and the New Century Excellent Talents in University Program of Ministry of Education of China under Grant No. NCET-10-0029, the Fundamental Research Funds for the Central Universities, and the Research Fund for the Doctoral Program of Higher Education under Grant No. 20110001110087.

References

1. S. Durr, et al., Science **322**, 1224 (2008).
2. ETM Collab. (C. Alexandrou et al.), Phys. Rev. D **80**, 114503 (2009).
3. PACS-CS Collab. (S. Aoki et al.), Phys. Rev. D **79**, 034503 (2009).
4. PACS-CS Collab. (S. Aoki et al.), Phys. Rev. D **81**, 074503 (2010).
5. A. Walker-Loud, et al., Phys. Rev. D **79**, 054502 (2009).
6. Hadron Spectrum Collab. (Huey-Wen Lin et al.), Phys. Rev. D **79**, 034502 (2009).
7. W. Bietenholz, et al., Phys. Lett. B **690**, 436 (2010).

8. W. Bietenholz, *et al.*, *Phys. Rev. D* **84**, 054509 (2011).

9. S.R. Beane, *et al.*, *Phys. Rev. D* **84**, 014507 (2011).

10. S. Weinberg, *Physica A* **96**, 327 (1979).

11. D. B. Leinweber, A. W. Thomas, and R. D. Young, *Phys. Rev. Lett.* **92**, 242002 (2004).

12. V. Bernard, T. R. Hemmert, and U.-G. Meissner, *Nucl. Phys. A* **732**, 149 (2004).

13. M. Procura, T. R. Hemmert, and W. Weise, *Phys. Rev. D* **69**, 034505 (2004).

14. V. Bernard, T. R. Hemmert, and U.-G. Meissner, *Phys. Lett. B* **622**, 141 (2005).

15. J. Gasser and H. Leutwyler, *Phys. Lett. B* **184**, 83 (1987).

16. J. Gasser and H. Leutwyler, *Nucl. Phys. B* **307**, 779 (1988).

17. S. R. Beane and M. J. Savage, *Phys. Rev. D* **68**, 114502 (2003).

18. D. Arndt and B. C. Tiburzi, *Phys. Rev. D* **69**, 114503 (2004).

19. X.-L. Ren, L.S. Geng, J. Martin Camalich, J. Meng, and H. Toki, *J. High Energy Phys.* **12**, 073 (2012).

20. X.-L. Ren, L.S. Geng, J. Meng, and H. Toki, *Phys. Rev. D* **87**, 074001 (2013).

21. K. Symanzik, *Nucl. Phys. B* **226**, 187 (1983), *Nucl. Phys. B* **226**, 205 (1983).

22. B. Sheikholeslami and R. Wohlert, *Nucl. Phys. B* **259**, 572 (1985).

23. X. -L. Ren, L. -S. Geng and J. Meng, *Eur. Phys. J. C* **74**, 2754 (2014).

24. X. -L. Ren, L. -S. Geng and J. Meng, arXiv:1404.4799 [hep-ph].

25. X. -L. Ren, L. -S. Geng and J. Meng, *Phys. Rev. D* **89**, 054034 (2014).

Roy–Steiner equations for πN scattering*

J. Ruiz de Elvira[a,†], C. Ditsche[a], M. Hoferichter[b], B. Kubis[a], U.-G. Meißner[a,c]

[a] Helmholtz-Institut für Strahlen- und Kernphysik (Theorie) and
Bethe Center for Theoretical Physics, Universität Bonn, D–53115 Bonn, Germany
[b] Albert Einstein Center for Fundamental Physics, Institute for Theoretical Physics,
Universität Bern, CH–3012 Bern, Switzerland
[c] Institut für Kernphysik, Institute for Advanced Simulation, and
Jülich Center for Hadron Physics, Forschungszentrum Jülich, D–52425 Jülich,
Germany

In this talk, we briefly review our ongoing collaboration to precisely determine
the low-energy πN scattering amplitude by means of Roy–Steiner equations.
After giving a brief overview of this system of dispersive equations and their
application to πN scattering, we proceed to solve for the lower partial waves
of the s-channel ($\pi N \to \pi N$) and the t-channel ($\pi\pi \to \bar{N}N$) sub-problems.

1. Introduction

Pion–nucleon scattering is one of the most fundamental processes involving the lightest mesons and baryons, allowing for a test of the dynamical constraints imposed by chiral symmetry at low energies. However, despite numerous investigations performed, the πN scattering amplitude is still not known to sufficient precision in the low-energy region. This is particularly striking in the scalar-isoscalar sector, where the determination of the pion–nucleon σ-term is still far from satisfactory.

Dispersion relations have repeatedly proven to be a powerful tool for studying processes at low energies with high precision.[1–4] They are built upon very general principles such as Lorentz invariance, unitarity, crossing symmetry, and analyticity. Moreover, the dispersive formalism is model independent and relates amplitudes at a given energy with an integral over the whole energy range, increasing the precision and providing information on the amplitude either at energies where data are poor.

*This research was supported by the DFG (SFB/TR 16).
†Speaker.

In particular, for $\pi\pi$ scattering, Roy equations (7) are obtained from a twice-subtracted fixed-t dispersion relation, where the t-dependent subtraction constants are determined by means of $s \leftrightarrow t$ crossing symmetry, and performing a partial-wave expansion. This leads to a coupled system of partial-wave dispersion relations (PWDRs), where the scattering lengths—the only free parameters—appear in the subtraction terms.

Unfortunately, in the case of πN scattering, a full system of PWDRs has to include dispersion relations for two distinct physical processes, $\pi N \to \pi N$ (s-channel) and $\pi\pi \to \bar{N}N$ (t-channel), and the use of $s \leftrightarrow t$ crossing symmetry will intertwine s- and t-channel equations.

2. Roy–Steiner Equations

Roy–Steiner (RS) equations (5) solve this problem by combining the s- and t- channel physical region by means of hyperbolic dispersion relations (HDRs). They are obtained by expanding the absorptive part of the HDRs into s- and t-channel partial waves, respectively, and subsequently by projecting the full HDRs onto s-channel partial waves, conventionally denoted by $f_{l\pm}^I$ with isospin index $I \in \{+, -\}$ and total angular momentum $j = l \pm 1/2 = l\pm$, or t-channel partial waves $f_\pm^J$, labeled by the parallel $(+)$/anti-parallel $(-)$ antinucleon–nucleon helicities and the total t-channel angular momentum.

Therefore, the s-channel RS equations read[5]

$$f_{l+}^I(W) = N_{l+}^I(W) + \frac{1}{\pi} \int\limits_{t_\pi}^{\infty} dt' \sum_J \left\{ G_{lJ}(W, t') \operatorname{Im} f_+^J(t') + H_{lJ}(W, t') \operatorname{Im} f_-^J(t') \right\} \quad (1)$$

$$+ \frac{1}{\pi} \int\limits_{W_+}^{\infty} dW' \sum_{l'=0}^{\infty} \left\{ K_{ll'}^I(W, W') \operatorname{Im} f_{l'+}^I(W') + K_{ll'}^I(W, -W') \operatorname{Im} f_{(l'+1)-}^I(W') \right\},$$

where due to G-parity only even/odd J contribute for isospin $I = +/-$, respectively. The kernels $K_{ll'}^I(W, W)$, $G_{lJ}(W, t)$, and $H_{lJ}(W, t)$ are known analytically, and $N_{l+}^I(W)$ denotes the partial-wave projections of the pole terms.

For the t-channel partial-wave projection, the corresponding t-channel RS equations are given by[6]

$$f_+^J(t) = \tilde{N}_+^J(t) + \frac{1}{\pi} \int\limits_{t_\pi}^{\infty} dt' \sum_{J'} \left\{ \tilde{K}_{JJ'}^1(t, t') \operatorname{Im} f_+^{J'}(t') + \tilde{K}_{JJ'}^2(t, t') \operatorname{Im} f_-^{J'}(t') \right\}$$

$$+ \frac{1}{\pi} \int\limits_{W_+}^{\infty} dW' \sum_{\ell=0}^{\infty} \left\{ \tilde{G}_{J\ell}(t, W') \operatorname{Im} f_{\ell+}^I(W') + \tilde{G}_{J\ell}(t, -W') \operatorname{Im} f_{(\ell+1)-}^I(W') \right\}, \quad (2)$$

and similarly for f_-^J except for the fact that these do not receive contributions from f_+^J. In addition, only even or odd J' couple to even or odd J (corresponding to t-channel isospin $I_t = 0$ or $I_t = 1$), respectively, and only higher t-channel partial waves contribute to lower ones.

3. Solutions of the t-Channel and s-Channel Subproblems

Due to its simpler recoupling scheme, the t-channel RS subsystem can be recast as a Muskhelishvili–Omnès (MO) problem[8] with a finite matching point,[3] where the inhomogeneities subsume the nucleon pole terms, all s-channel integrals, and the higher t-channel partial waves.[6]

The numerical solutions for the lowest partial waves of this MO problem were studied in Ref. 6 using the KH80 analysis[9] as input. For the P-waves, a single-channel approximation was considered, whereas for the S-waves a two-channel description including $\bar{K}K$ intermediate states was used. We plot in Fig. 1 the solution for the real part of $f_+^0(t)$ and the absolute value of $f_+^1(t)$, which show a nice consistency with the KH80 results.

Once the t-channel equations are solved, the structure of the s-channel problem resembles the form of $\pi\pi$ Roy equations, and should be amenable to similar solution techniques.

As a first step, in order to investigate to what extent these equations are fulfilled for the SAID s-channel amplitudes,[10] we compare the left- and right-hand side of Eq. (1). In Fig. 2, we present the results for the S_{11} and P_{13} partial waves, which show that the equation are fulfilled in the threshold region, while deviations emerge at higher energies in both waves.

In order to improve these results, we impose as a constraint the modern πN coupling and S-wave scattering lengths determinations of Ref. 11, ob-

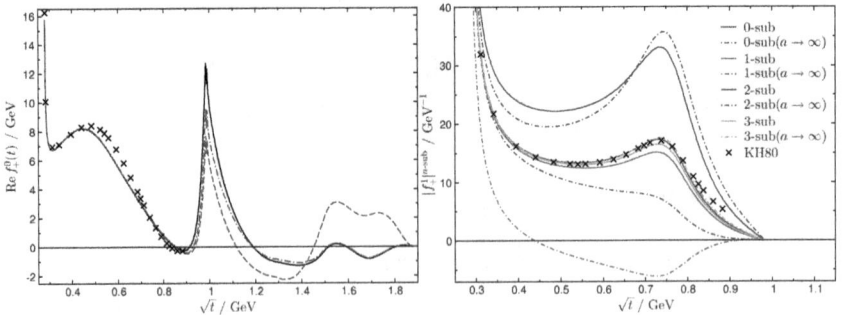

Fig. 1. Left: result for $\mathrm{Re} f_+^0(t)$. The solid, dashed, and dot-dashed lines refer to three different variants of the input above the matching point.[6] The black crosses indicate the KH80 results.[9] Right: solution for $|f_+^1(t)|$. The different lines correspond to the number of subtractions considered.[6]

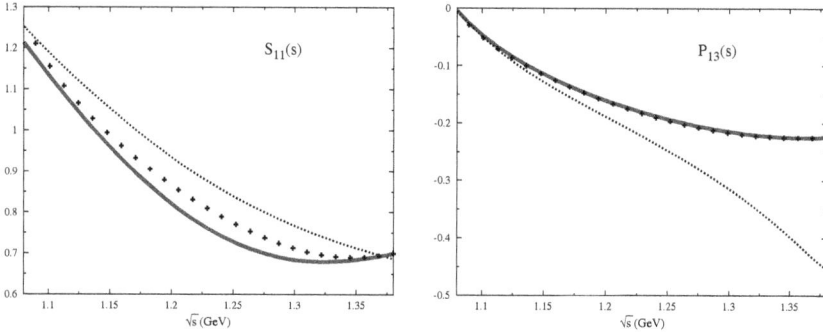

Fig. 2. RS results for the real part of the S_{11} and P_{13} partial waves. The black crosses indicate the SAID results,[10] i.e. the left-hand side of Eq. (1) before the fit, whereas the solid line corresponds to the left-hand side of Eq. (1) after the fit. The dotted and dashed lines denote the right-hand side of Eq. (1) before and after the fit, respectively.

tained from hadronic-atom data, and minimize the difference between the left- and right-hand side of Eq. (1), by fitting the s-channel phase shifts together with the subtraction constants. The fitted results are also illustrated in Fig. 2, showing that the agreement between left- and right-hand side is now very good. For the P_{13} wave, this agreement is mainly due to the change of the subthreshold parameters, since the difference between the left-hand side of the RS equations before and after the fit is negligible, while for the S_{11} wave also the partial wave changes significantly.

The next step of this project will be a self-consistent iteration procedure between the solutions for the s- and t-channel, leading to a consistent and precise description of the low-energy πN scattering amplitude.

References

1. B. Ananthanarayan et al., Phys. Rept. **353** (2001) 207.
2. R. García-Martín et al., Phys. Rev. D **83** (2011) 074004.
3. P. Büttiker et al., Eur. Phys. J. C **33** (2004) 409.
4. M. Hoferichter, D. R. Phillips and C. Schat, Eur. Phys. J. C **71** (2011) 1743.
5. G. E. Hite and F. Steiner, Nuovo Cim. A **18** (1973) 237.
6. C. Ditsche et al., JHEP **1206** (2012) 043; JHEP **1206** (2012) 063.
7. S. M. Roy, Phys. Lett. B **36** (1971) 353.
8. N. I. Muskhelishvili, *Singular Integral Equations*, Wolters-Noordhoff Publishing, Groningen, 1953; R. Omnès, Nuovo Cim. **8** (1958) 316.
9. R. Koch and E. Pietarinen, Nucl. Phys. A **336** (1980) 331.
10. R. A. Arndt et al., Eur. Phys. J. A **35** (2008) 311.
11. V. Baru *et al.*, Phys. Lett. B **694** (2011) 473; Nucl. Phys. A **872** (2011) 69.

Experimental results on $Z_c(3900)$

C. P. Shen

School of Physics and Nuclear Energy Engineering, Beihang University
Beijing, 100191, China
E-mail:shencp@ihep.ac.cn

This report reviewed the recently discovered $Z_c(3900)$ at around 3.9 GeV/c^2 in the $\pi^{\pm}J/\psi$ mass spectrum by the Belle and BESIII collaborations simultaneously. Belle collaboration observed it in the process $e^+e^- \to \pi^+\pi^- J/\psi$ within the $Y(4260)$ signal region with a 967 fb^{-1} data sample using initial-state-radiation technology. BESIII collaboration discovered it in the same process at a fixed center-of-mass energy of 4.260 GeV using a 525 pb^{-1} data sample. The measured resonance masses and widths from Belle and BESIII measurements are consistent with each other within the errors. The $Z_c(3900)$ can be interpreted as a new charged charmonium-like state.

Keywords: $Z_c(3900)$; $Y(4260)$; charged charmonium-like state.

1. Introduction

As we know, currently QCD is regarded as a fundamental theory of strong interactions. Systems that include heavy quark-antiquark pairs (quarkonia) are a ideal laboratory for probing both the high energy regime of QCD and the low energy regime. The accuracy of current models of quarkonia is such that a particle which mimics quarkonia but does not fit in the model spectrum is a likely candidate for a nonconventional, "exotic" state (XYZ particles).

Since the $X(3872)$ was discovered by Belle ten years ago, lots of charmonium-like resonances have been observed in the final states with a charmonium and some light hadrons. They could be candidates for usual charmonium states, however, there are also lots of strange properties shown from these states. It is not at all clear what most of the new XYZ states are.

Motivated by the striking observations of charged charmonium-like[1,2] and bottomonium-like states,[3] Belle and BESIII investigate the existence of similar states as intermediate resonances in $Y(4260) \to \pi^+\pi^- J/\psi$ decays.[4,5]

2. Discovery of the $Z_c(3900)$

After all the event selections, Figure 1 shows the Dalitz plots of events in the J/ψ signal region from Belle and BESIII data, where there are structures in the $\pi^+\pi^-$ system and evidence for an exotic charmoniumlike structure in the $\pi^\pm J/\psi$ system.

Figures 2 and 3 show the projections of the $M(\pi^+\pi^-)$, $M(\pi^+ J/\psi)$, and $M(\pi^- J/\psi)$ distributions for the signal events, as well as the background events estimated from normalized J/ψ mass sidebands. In the $\pi^\pm J/\psi$ mass spectrum, there is a significant peak at around 3.9 GeV/c^2 (denoted as the $Z_c(3900)$). The wider peak at low mass is a reflection of the $Z_c(3900)$ as

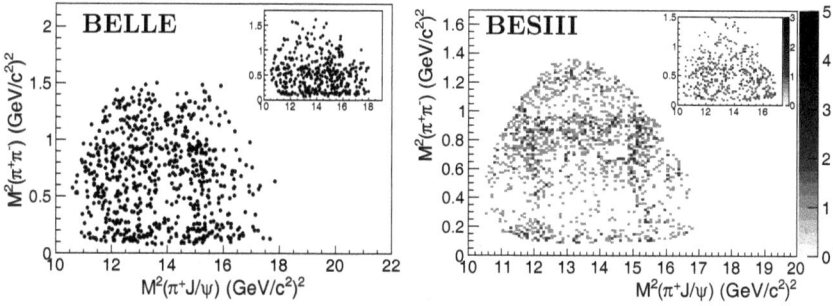

Fig. 1. Dalitz distributions of $M^2(\pi^+\pi^-)$ versus $M^2(\pi^+ J/\psi)$ for selected $\pi^+\pi^- J/\psi$ events in the J/ψ signal region. The inset shows background events from the J/ψ-mass sidebands (not normalized).

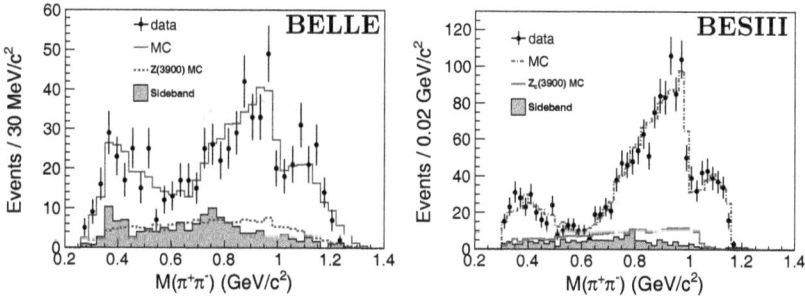

Fig. 2. Invariant mass distributions of $\pi^+\pi^-$ for events in the J/ψ signal region from Belle and BESIII experimental data. Points with error bars represent data, shaded histograms are normalized background estimates from the J/ψ-mass sidebands, red histograms represent MC simulation results from $\sigma(500)$, $f_0(980)$ and non-resonant $\pi^+\pi^-$ amplitudes and lower histograms are MC simulation results for a $Z(3900)^\pm$ signal.

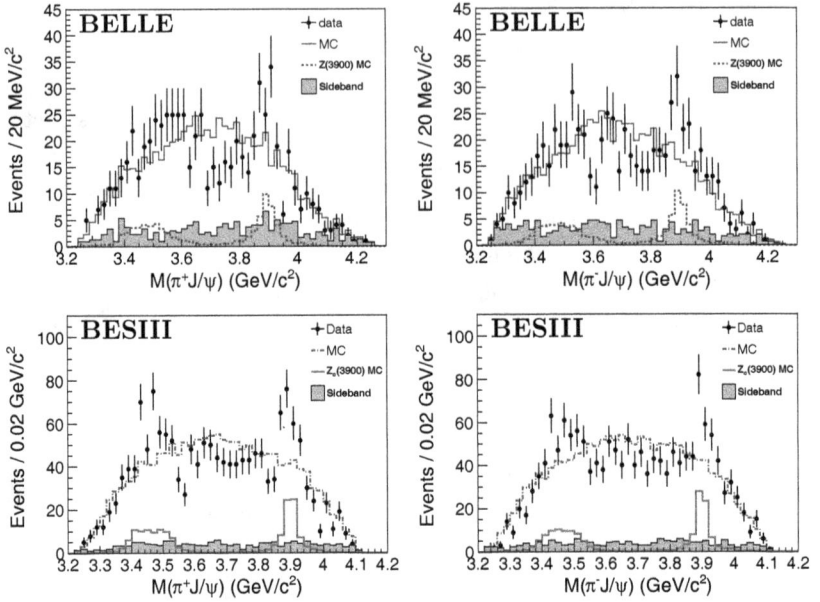

Fig. 3. One dimensional projections of the $M(\pi^+ J/\psi)$ and $M(\pi^- J/\psi)$ for events in the J/ψ signal region from Belle and BESIII experimental data. Points with error bars represent data, shaded histograms are normalized background estimates from the J/ψ-mass sidebands, red histograms represent MC simulation results from $\sigma(500)$, $f_0(980)$ and non-resonant $\pi^+\pi^-$ amplitudes and lower histograms are MC simulation results for a $Z(3900)^\pm$ signal.

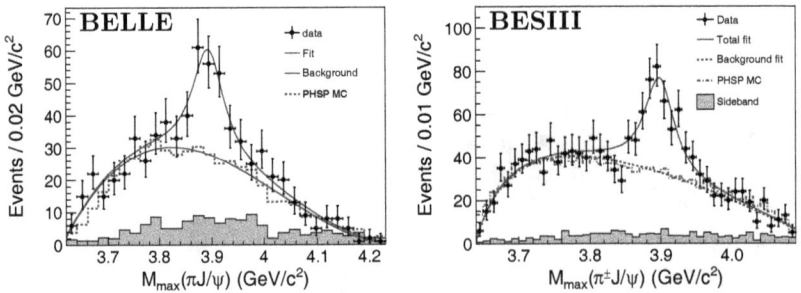

Fig. 4. Unbinned maximum likelihood fits to the distributions of the $M_{\max}(\pi J/\psi)$ from Belle and BESIII experimental data. Dots with error bars are data, the solid curves are the best fits, the dashed histograms represent the results of phase space distribution and the shaded histograms are the normalized J/ψ sideband events.

indicated from MC simulation. The $\pi^+\pi^-$ mass spectrum shows complicated structures. A parameterization for the $\pi^+\pi^-$ mass spectrum that includes a $f_0(980)$, $\sigma(500)$, and a non-resonant amplitude can describe the data well, but does not generate any peaking structure in the $\pi^\pm J/\psi$ projection.

Unbinned maximum likelihood fits are applied to the distributions of $M_{\max}(\pi^\pm J/\psi)$ from Belle and BESIII data. The signal shape is parameterized as an S-wave Breit-Wigner function convolved with a Gaussian with a mass resolution fixed at the MC simulated value. The mass-dependent efficiency is also included in the fit and the possible interference between the signal and background is neglected. Figure 4 shows the fit results. The measured masses are $(3899.0\pm3.6\pm4.9)\,\mathrm{MeV}/c^2$ and $(3894.5\pm6.6\pm4.5)\,\mathrm{MeV}/c^2$ and the measured widths are $(46\pm10\pm20)\,\mathrm{MeV}$ and $(63\pm24\pm26)\,\mathrm{MeV}/c^2$ from Belle and BESIII experiments, respectively. They are consistent each other within the errors. The signal significance is greater than 5σ in both of the measurements. This state is close to the $D\bar{D}^*$ mass threshold. As the $Z(3900)^\pm$ state has a strong coupling to charmonium and is charged, it cannot be a conventional $c\bar{c}$ state.

3. Summary and Discussion

The $Z_c(3900)$ has been observed clearly in the $\pi^\pm J/\psi$ mass spectrum by Belle and BESIII collaborations in the same process $Y(4260) \to \pi^+\pi^- J/\psi$ at the same time. Since the observation of the $Z_c(3900)$, there have been a number of different interpretations, including tetraquark state, hadronic molecule, hadron-charmonium state and so on. To understand its nature, more efforts in both of experiments and theory are needed.

Acknowledgments

This work is supported partly by the Fundamental Research Funds for the Central Universities of China (303236).

References

1. S. K. Choi *et al.* (Belle Collaboration), Phys. Rev. Lett. **100**, 142001 (2008).
2. R. Mizuk *et al.* (Belle Collaboration), Phys. Rev. D **78**, 072004 (2008).
3. A. Bondar *et al.* (Belle Collaboration), Phys. Rev. Lett. **108**, 122001 (2012).
4. Z. Q. Liu *et al.* (Belle Collaboration), Phys. Rev. Lett. **110**, 252002 (2013).
5. M. Ablikim *et al.* (BESIII Collaboration), Phys. Rev. Lett. **110**, 252001 (2013).

Dynamically generated resonances from the vector octet-baryon octet interaction with strangeness zero

B.-X. Sun*, J. Wang

College of Applied Sciences, Beijing University of Technology, Beijing 100124, China
* E-mail: sunbx@bjut.edu.cn

X.-F. Lu

Department of Physics, Sichuan University, Chengdu 610065, China

The interaction potentials between vector mesons and baryon octet are calculated explicitly with a summation of $t-$, $s-$, $u-$ channel diagrams and a
contact term originating from the tensor interaction. Altogether, 17 resonances
are generated dynamically in different channels of strangeness zero by solving
the relativistic Lippman-Schwinger equations in the S-wave approximation of
partial wave analysis, and their masses, decay widths, isospins and spins are
determined. Some resonances are well fitted with their counterparts listed in
the newest review of Particle Data Group(PDG),[1] while others might stimulate
the experimental observation in these energy regions in the future.

Keywords: Hadronic resonances; Hidden gauge symmetry; coupled-channel
unitary approach.

The nonperturbative unitary techniques, which are combined with the chiral Lagrangian of mesons and baryons, have been used to study the meson-
meson and meson-baryon interactions and new states in the resonance
region, which are not easily explained using the conventional constituent
quark model. Moreover, the theory on the hidden gauge symmetry supplies
a mechanism to include the vector meson in the chiral Lagrangian.[2,3] Along
this clue, the interactions between the vector meson and other mesons have
been studied and some resonances are generated dynamically.[4–6]

In the baryon sector, the interactions of the vector meson with baryon
decuplet and octet are addressed respectively in Refs.,[7–9] where only $t-$
channel amplitudes are analyzed in the S-wave approximation. In Ref.,[10]
the $t-$,$s-$, $u-$ channel diagrams and a contact diagram originating from

the tensor term of the vector meson and baryon octet interaction are all taken into account, and four spin-determined resonances are found in a non-relativistic approximation. In this work, we deduce the interaction kernel of vector mesons and baryon octet including a vector meson exchange in $t-$ channel, octet baryon exchange in $s-$, $u-$ channels, and a contact diagram related only to the tensor interaction term precisely, and then calculate the scattering amplitudes of vector mesons and baryon octet by solving the coupled channel Lippman-Schwinger equation in the framework of partial wave analysis, where only the S-wave of the initial vector meson is taken into account.

We follow the formalism of the vector meson-baryon octet interaction in Ref.,[10] and then the Lagrangian for the linearly coupling to vector mesons can be explicitly written as

$$
\begin{aligned}
\mathcal{L}_{VBB} = -g\Big\{ &F_V \langle \bar{B}\gamma_\mu \left[V_8^\mu, B \right] \rangle + D_V \langle \bar{B}\gamma_\mu \{ V_8^\mu, B \} \rangle \\
&+ \frac{1}{4M}\Big(F_T \langle \bar{B}\sigma_{\mu\nu} \left[\partial^\mu V_8^\nu - \partial^\nu V_8^\mu, B \right] \rangle + D_T \langle \bar{B}\sigma_{\mu\nu} \{ \partial^\mu V_8^\nu - \partial^\nu V_8^\mu, B \} \rangle \Big) \\
&+ \frac{C_V}{3} \langle \bar{B}\gamma_\mu B \rangle \langle V_0^\mu \rangle + \frac{C_T}{4M} \langle \bar{B}\sigma_{\mu\nu} V_0^{\mu\nu} B \rangle \Big\}
\end{aligned}
\tag{1}
$$

and

$$
\mathcal{L}_{VVBB} = \frac{g}{4M}\Big\{ F_T \langle \bar{B}\sigma_{\mu\nu} \left[ig\left[V_8^\mu, V_8^\nu \right], B \right] \rangle + D_T \langle \bar{B}\sigma_{\mu\nu} \{ ig\left[V_8^\mu, V_8^\nu \right], B \} \rangle \Big\}.
\tag{2}
$$

From Eqs. (1) and (2), we can obtain the transition potentials for the $t-$, $s-$, $u-$ channels and contact interactions between baryons and vector mesons.

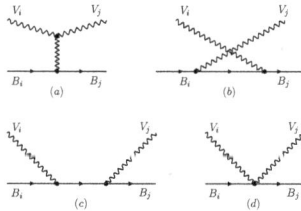

Fig. 1.   Feynman Diagrams of the vector meson-baryon interaction. (a) $t-$ channel, (b) $u-$ channel, (c) $s-$ channel, and (d) contact term.

The scattering matrix implies solving the relativistic coupling-channel Lippman-Schwinger equation in the on-shell factorization approach

$$
T(\sqrt{s}, \cos\theta) = [1 - V(\sqrt{s}, \cos\theta) G]^{-1} V(\sqrt{s}, \cos\theta)
\tag{3}
$$

with G being the loop function of a vector meson and a baryon.

Table 1. The experimental values on the coupling constants of vector mesons to the baryons.[11]

$Exp.$	ω	ρ	K^*	ϕ
$g^2/4\pi$	2.4	2.4	1.39	12.0

The experimental values on coupling constants of vector mesons to baryons are listed in Table 1, which are taken from Ref.[11] Thus we can fit the parameters F_V, D_V and C_V in the Lagrangian with these data. Moreover, we assume the parameters F_T, D_T and C_T take the same values as F_V, D_V and C_V, respectively. The values of these parameters are listed in Table 2.

In the s-wave approximation of partial wave analysis, only the final state with the orbital angular momentum $l' = 0$ is studied. Since the parity is conserved, the contribution from the initial state with the orbital angular momentum $l = 0, 2$ must be taken into account. Thus the scattering am-

Table 2. The parameters used in the calculation.

F_V	D_V	F_T	D_T	C_V	C_T
1.6405	0.2225	1.6405	0.2225	-5.144	-5.144

Table 3. The properties of the dynamically generated resonance with isospin $I = 1/2$ and its possible PDG counterpart.

J	Theory	PDG data			
	Pole positions	Name and J^P	Status	Mass	Width
1/2	$1648 + i0$ MeV	$N(1650)1/2^-$	****	$1645 - 1670$ MeV	$120 - 180$ MeV
1/2	$1705 + i0$ MeV				
1/2	$1771 + i0$ MeV				
1/2	$1846 + i10$ MeV	$N(1895)1/2^-$	**	≈ 2090 MeV	$100 - 400$ MeV
3/2	$1630 + i30$ MeV				
3/2	$1669 + i5$ MeV				
3/2	$1687 + i0$ MeV	$N(1685)?^?$	*	$1670 - 1685$ MeV	≈ 25 MeV
3/2	$1729 + i0$ MeV	$N(1700)3/2^-$	***	$1650 - 1750$ MeV	$100 - 250$ MeV
3/2	$1870 + i20$ MeV	$N(1875)3/2^-$	***	$1820 - 1920$ MeV	$160 - 320$ MeV
3/2	$1954 + i30$ MeV				
3/2	$2035 + i5$ MeV	$N(2120)3/2^-$	**	≈ 2120 MeV	≈ 300 MeV

Table 4. The properties of the dynamically generated resonance with isospin $I = 3/2$ and its possible PDG counterpart.

I, I_z	J	Theory	PDG data			
		Pole positions	Name and J^P	Status	Mass	Width
$3/2, 3/2$	$1/2$	$1642 + i5$ MeV	$\Delta(1620)1/2^-$	****	$1600 - 1660$ MeV	$130 - 150$ MeV
$3/2, 3/2$	$1/2$	$1825 + i5$ MeV				
$3/2, 3/2$	$1/2$	$1855 + i5$ MeV				
$3/2, 3/2$	$3/2$	$1738 + i5$ MeV	$\Delta(1700)3/2^-$	****	$1670 - 1750$ MeV	$200 - 400$ MeV
$3/2, 1/2$	$1/2$	$1930 + i5$ MeV	$\Delta(1900)1/2^-$	**	$1840 - 1920$ MeV	$140 - 300$ MeV
$3/2, 1/2$	$3/2$	$1960 + i5$ MeV	$\Delta(1940)3/2^-$	**	$1940 - 2060$ MeV	$200 - 460$ MeV

plitude can be expanded in terms of amplitudes with fixed total angular momenta J.

For the case of isospin $I = 1/2$, four resonances are generated dynamically in the channel of $J = 1/2$ and seven resonances in the channel of $J = 3/2$ the pole positions in the complex energy plane and the possible counterparts are listed in Table 3. The channels of isospin $I = 3/2$ are also studied and the pole positions are summarized in Table 4. In addition to the resonances $\Delta(1620)1/2^-$ and $\Delta(1700)3/2^-$, there are other two poles produced above $1900 MeV$ in the channel of isospin $I = 3/2$ and $I_z = 1/2$, which might correspond to the resonance $\Delta(1900)1/2^-$ and $\Delta(1940)3/2^-$ although these two particles have been omitted in the latest PDG data.

References

1. J. Beringer et al. (Particle Data Group), Phys. Rev. D86, 010001 (2012).
2. M. Bando, T. Kugo, S. Uehara, K. Yamawaki and T. Yanagida, Phys. Rev. Lett. **54**, 1215 (1985).
3. U. G. Meissner, Phys. Rep. **161**, 213 (1988).
4. H. Nagahiro, L. Roca, A. Hosaka and E. Oset, Phys. Rev. D **79**, 014015 (2009).
5. R. Molina, D. Nicmorus and E. Oset, Phys. Rev. D **78**, 114018 (2008).
6. L. Geng and E. Oset, Phys. Rev. D **79**, 074009 (2009).
7. P. Gonzalez, E. Oset and J. Vijande, Phys. Rev. C **79**, 025209 (2009).
8. S. Sarkar, B. -X. Sun, E. Oset, M. J. Vicente Vacas, Eur. Phys. J. A **44**, 431-443 (2010).
9. E. Oset and A. Ramos, Eur. Phys. J. A **44**, 445 (2010).
10. K. P. Khemchandani, H. Kaneko, H. Nagahiro and A. Hosaka, Phys. Rev. D83, 114041 (2011).
11. L. J. Reinders, H. Rubinstein and S. Yazaki, Phys. Rept. 127, 1 (1985).

Massive hybrid stars with strangeness

T. Takatsuka* and T. Hatsuda

Theoretical Research Division, Nishina Center, RIKEN, Wako 351-0198, Japan
* E-mail: takatuka@iwate-u.ac.jp

K. Masuda

Department of Physics, The University of Tokyo, Tokyo 113-0033, Japan

How massive the hybrid stars could be is discussed by a "3-window model"proposed from a new strategy to construct the equation of state with hadron-quark transition. It is found that hybrid stars have a strong potentiality to generate a large mass compatible with two-solar-mass neutron star observations.

1. Introduction

It seems a recent consensus that hyperons (Y) are sure to participate in neutron star (NS) cores, increasing the population with the increase of baryon density (ρ).[1] The Y-mixing, as a manifestation of strangeness degrees of freedom, plays a dramatic role in NS properties, that is, it causes an extreme softening of the EOS,[2-8] leading to the problem that the maximum mass (M_{max}) of NSs cannot exceed even the 1.44 $M_\odot$ observed for PSR1913+16. This conflict between the theory and the observation becomes more serious by a very recent finding of $2M_\odot$-NSs.[9,10] In a pure hadronic framework, it has been pointed out that the introduction of a "universal 3-body force"acting on all the baryons BBB (i.e., not only on NNN but also on NNY, NYY and YYY) is a promising candidate to solve the problem.[11]

The aim of this paper is to discuss another solution for the problem by extending the framework from pure hadron to hadron (H) plus quark (Q) degrees of freedom. We address how the hybrid stars with H-Q transition core could be massive, by a new approach not restricted to the conventional Gibbs or Maxwell condition. Our new strategy is to divide the equation of state (EOS) into 3 density regions, i.e., pure H-EOS for $\rho \leq \rho_H$, HQ-EOS for $\rho_H \leq \rho \leq \rho_Q$ and pure Q-EOS for $\rho \geq \rho_Q$, characterized as "3-window model".[12] The motivation comes from the considerations: (i) Pure hadronic EOS gets uncertain with increasing ρ because of finite size hadrons composed of quarks and gluons. (ii) Pure quark matter EOS becomes unreliable with decreasing ρ due to the deconfined-confined transition. (iii) Therefore, to discuss the H-Q transition by extrapolating the pure H-EOS from a lower density side and the pure Q-EOS from a higher density side is not necessarily justified. Our basic idea is to supplement the very poorly known HQ-EOS by sandwitching it in between the relatively certain H-EOS and Q-EOS, and construct the HQ-EOS by a phenomenological interpolation.

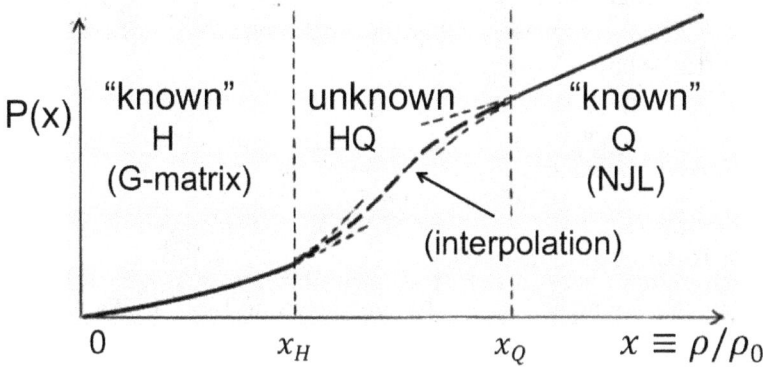

Fig. 1. Schematic illustration of "3-windouw model". A very poorly known HQ-EOS is interpolated by sandwiching it in between a H-EOS and a Q-EOS relatively known.

2. Approach

According to the "3-window model", we construct the EOS with H-Q transition. In our preceding works,[13,14] we have tried this line of approach from a view of a smooth crossover for the H-Q transition region and found that the hybrid stars satisfy $M_{max} \geq 2M_\odot$. There the pressure $P(\rho)$ was interpolated as $P(\rho) = P_H(\rho)f_-(\rho) + P_Q(\rho)f_+(\rho)$ by a ρ-dependent weight function $f_\pm(\rho) = (1\pm \tanh{[(\rho - \bar{\rho})/\Gamma]})$ with parameters $\bar{\rho}$ and Γ, in an analogy to very hot QCD transition. Due to $f_\pm(\rho)$, however, the interpolated HQ-EOS approaches only asymptotically to H-EOS with decreasing ρ (Q-EOS with increasing ρ). But such a way of interpolation is not unique. As a complementary work, here we try more general interpolation and make exact matching at discrete boundaries (ρ_H and ρ_Q).

As in the preceding work,[14] we take the H-EOS with Y (denoted by TNI2) from a G-matrix effective interaction approach. The TNI2 H-EOS satisfies the saturation property of symmetric nuclear matter and has an incompressibility $\kappa = 250\text{MeV}$ consistent with experiments. We use the 3-flavor Q-EOS from the NJL model including a repulsive effect from vector interaction with the strength $g_v = (0 - 1.5)G_s$ (G_s being the strength of scalar interaction). As an interpolation function, we take $P_{HQ}(x) = ax^m + bx^n + c$ with $x \equiv \rho/\rho_0$ ($\rho_0 = 0.17/\text{fm}^3$ being the nuclear density). Then, the energy density ϵ is obtained from $P = \rho^2 \partial(\epsilon/\rho)/\partial\rho$ as $\epsilon_{HQ}(x) = (a/(m-1))x^m + (b/(n-1))x^n + dx - c$. Four coeficients $\{a, b, c, d\}$ are determined for a given set $\{m, n\}$ and $\{x_H \equiv \rho_H/\rho_0, x_Q \equiv \rho_Q/\rho_0\}$ by a matching of P and ϵ at phase boundaries. By running the set of $\{m, n\}$ and $\{x_H, x_Q\}$, the solution is searched under the conditions; (i) $P(x) > 0$ and $\partial P/\partial x \geq 0$ (thermodynamic stability), (ii) $v_s/c = (\partial P/\partial\epsilon)^{1/2} \leq 1$ (sound velocity less than light velocity), (iii) $x_H > 1$ (no experimental evidence for quark degrees of freedom at $\rho \leq \rho_0$).

3. Some Results and Remarks

Some examples for numerical results are shown in Table 1. We note the following points: (i) Within the present interpolation function, we have several hybrid stars with $M_{max} \simeq (2-3)M_\odot$. It can be as massive as $3M_\odot$-NSs. (ii) The dependence of M_{max} on $\{m\ n\}$ and $\{x_H, x_Q\}$ is rather small: For a fixed $\{m=0.2, n=-2.6, x_H = 1.5\}$ and $g_v = 0.5G_s$, M_{max} changes slightly, $(2.61 \rightarrow 2.48)M_\odot$ according to $x_Q = (5.5 \rightarrow 8.0)$. For a fixed $\{x_H = 1.5, x_Q = 0.7\}$ and $g_v = 0.5G_s$, the functional dependence of M_{max} is also small as $M_{max}=(2.53, 2.62, 2.61)M_\odot$ for (m, n)=(0.2, -2.6), (2.6, -0.2), (1.2, -1.2). (iii) The g_v-dependence of M_{max} is remarkable as

Table 1.   Some results for NS models

CASE	x_H	x_S	H-EOS	Q-EOS	m	n	$M_{max}/M_\odot$	R/km	ρ_c/ρ_0
1	1.5	5.5	TNI2	$g_v = 0.5G_s$	0.2	-2.6	2.61	13.38	3.99
2	1.5	6.0	TNI2	$g_v = 0.5G_s$	0.2	-2.6	2.59	13.27	3.90
3	1.5	7.0	TNI2	$g_v = 0.5G_s$	0.2	-2.6	2.53	12.08	4.52
4	1.5	8.0	TNI2	$g_v = 0.5G_s$	0.2	-2.6	2.48	12.56	4.35
5	1.5	7.0	TNI2	$g_v = 1.5G_s$	0.2	-2.6	3.08	13.73	3.34
6	1.5	7.0	TNI2	$g_v = 1.0G_s$	0.2	-2.6	2.86	13.28	3.94
7	1.5	7.0	TNI2	$g_v = 0$	0.2	-2.6	1.99	12.30	4.85
8	1.5	7.0	TNI2	$g_v = 0.5G_s$	2.6	-0.2	2.62	13.44	4.05
9	1.5	7.0	TNI2	$g_v = 0.5G_s$	1.2	-1.2	2.61	13.44	3.73

$M_{max}=(1.99,\ 2.53,\ 2.86,\ 3.08)M_\odot$ according to $g_v=(0.0,\ 0.5,\ 1.0,\ 1.5)G_s$ (CASE 7, 3, 6, 5). (iv) Since $x_Q > \rho_c$ (the central density), our hybrid stars do not have pure Q-matter core but H-Q transient core. In the calculations we have found that the x_H as lower as (1.5-2.5) is necessary for the solution to exist. This may suggest a picture that the Q-degrees of freedom begins to work at rather low density as has been discussed from a view of quark percolation in nuclear medium.[15]

To summarize, our hybrid stars from the "3-window model" can generate the M_{max} compatible with $2M_\odot$-NS observations, as far as the Q-degrees of freedowm sets on from a rather low density ($\sim 1.5\rho_0$) and the Q-EOS is stiff enough. The present work supports the results in our preceding papers. Finally, we want to stress that the quark degrees of freedom in NS cores has a potentiality enough to account for the existence of $2M_\odot$-NS.

Acknowledgments

This research was supported by JSPS Grant-in-Aid for Scientific Research (B) No.22340052 and by RIKEN 2012 Strategic Program for R&D.

References

1. T. Takatsuka, *Prog. Theor. Phys. Suppl.* **No.156**, 84 (2004) and references therein.
2. S. Nishizaki, Y. Yamamoto and T. Takatsuka, *Prog. Theor. Phys.* **105**, 607 (2001).
3. S. Nishizaki, Y. Yamamoto and T. Takatsuka, *Prog. Theor. Phys.* **108**. 703 (2002).
4. M. Baldo, G. F. Burgio and H. - J. Schulze, *Phys. Rev.* **C61**, 055801 (2000).
5. I. Vidaña, A. Polls, A. Ramos, L. Engvik and M. Hjorth-Jensen, *Phys. Rev.* **C62**, 035801 (2000).
6. Z. H. Li and H. - J. Schulze, *Phys. Rev.* **C78**, 028801 (2008).
7. K. Tsubakihara, H. Maekawa, H. Matsumiya and A. Ohnishi, *Phys. Rev.* **C81**, 065206 (2010).
8. H. Dapo, B. - J. Schaefer and J. Wambach, *Phys. Rev.* **C81**, 035803 (2010).
9. P. B. Demorest, et al., *Nature* **467**, 1081 (2010).
10. J. Antoniadis, et al., *Science* **340**, 6131 (2013).
11. T. Takatsuka, S. Nishizaki and R. Tamagaki, *Proc. Int. Symp. "FM50"* (AIP conference proceedings 1011) 209 (2008).
12. T. Takatsuka, T. Hatsuda and K. Masuda, *Proc. Int. Symp. "OMEG11"* (AIP conference proceedings 1484) 406 (2012).
13. K. Masuda, T. Hatsuda and T. Takatsuka, *Astrophys. J.* **794**, 12 (2013).
14. K. Masuda, T. Hatsuda and T. Takatsuka, *Prog. Theor. Exp. Phys.* 073D01 (2013).
15. G. Baym, *Physica* **96A**, 131 (1979).

Shape-phase transitions in very neutron-rich nuclei from $_{40}$Zr to $_{46}$Pd

Hiroshi Watanabe

IRCNPC, School of Physics and Nuclear Energy Engineering, Beihang University, Beijing 100191, China
E-mail: hiroshi@ribf.riken.jp

Gamma-ray spectroscopy following the β decay is an effective tool for exploring low-lying yrast and non-yrast states, which provide key structure information such as the shape transitions/coexistence and the single-particle levels. For the study of rare isotopes, especially when the nucleus of interest lies at the boundaries of availability for spectroscopic studies, isomeric decays are likely to be a more useful means than β decays to populate lower-lying levels. The identification of such characteristic isomers will pin down currently controversial subjects including the evolution of shell structures. The combined β- and isomeric-decay measurements at the RI Beam Factory (RIBF),[1] which has the capability of providing the world's strongest RI beams, are at the forefront of exploration of exotic nuclei far from the stability line.

This paper focuses on the achievements obtained in the first decay spectroscopy at RIBF in 2009. A major aim of this work was the study of neutron-rich nuclei around $Z = 40$ and $A = 110$, where the shape transitions from prolate, via γ-soft, to oblate deformations are predicted to occur with increasing the number of neutrons. New results obtained include the observation of shape evolution in ^{106,108}Zr,[2] a possible oblate-shape isomer in ^{109}Nb,[3] and a large-amplitude γ-soft dynamics in ^{110}Mo.[4] These findings revealed the shape-transitional phenomena in this neutron-rich region for the first time.

Experiments were carried out at the RIBF facility, cooperated by RIKEN Nishina Center and CNS, University of Tokyo. Neutron-rich $A \approx$ 110 nuclei were produced via in-flight fission of ^{238}U^{86+} projectiles at 345 MeV/nucleon, incident on a beryllium target with a thickness of 3 mm.

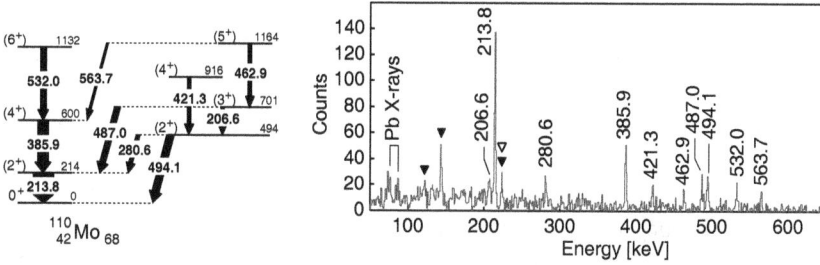

Fig. 1. Partial level scheme of ^{110}Mo (left) and γ-ray spectrum measured in coincidence with β rays detected within 250 ms after implantation of ^{110}Nb (right). Contaminants from ^{110}Tc and ^{109}Mo are indicated with filled and open triangles, respectively.

The average beam intensity was approximately 0.3 pnA during the experiment. The nuclei of interest were separated and transported through the BigRIPS spectrometer[5],[6] operated with a 6-mm-thick wedge-shaped aluminum degrader at the first dispersive focal plane for purification of the secondary beams. An additional degrader placed at the second dispersive focus served as a charge stripper to remove fragments that were not fully stripped. The identification of nuclei by their atomic number and the mass-to-charge ratio was achieved on the basis of the ΔE-TOF-$B\rho$ method, where ΔE, TOF, and $B\rho$ denote energy loss, time of flight, and magnetic rigidity, respectively. The identified nuclei were implanted into an active stopper consisting of nine double-sided silicon-strip detectors (DSSSD) stacked compactly. Each DSSSD has a thickness of 1 mm with a 50 mm × 50 mm active area segmented into sixteen strips on both sides in the vertical and horizontal dimensions. The DSSSDs also served as detectors for electrons following β-decay and internal-conversion processes. The implantation of an identified particle was associated with the subsequent electron events that were detected in the same DSSSD pixel. Gamma rays were detected by four Compton-suppressed Clover-type Ge detectors arranged around the DSSSD telescope in a close geometry. Further details of a particle-identification spectrum and data-analysis techniques are given in Ref.[3]

The left panel in Fig. 1 exhibits the level scheme of ^{110}Mo. Prior to the present work, the ground-state band in ^{110}Mo has been known up to the 10^+ state by measuring the prompt γ rays from the spontaneous fission of ^{248}Cm.[7] In addition to the 214-, 386-, and 532-keV γ rays that belong to the ground-state band, seven new transitions have been unambiguously observed in a singles γ-ray spectrum measured in coincidence with β rays subsequent to implantation of ^{110}Nb, as shown in the right panel in Fig. 1.

The second 2^+ level (2_2^+) is proposed at 494 keV, decaying by the 281- and 494-keV transitions which directly feed the yrast 2^+ and 0^+ states, respectively. The assignment of the 2_2^+ state at 494 keV is justified, since only the energy sum for the 281−214-keV cascade agrees with the energy of the 494-keV γ ray within experimental errors among the newly observed γ rays. The 207-keV transition is assigned as feeding the 2_2^+ state on account of the consistency in energy with the level at 701 keV, which is most likely the 3^+ member of the quasi-γ band decaying also via the parallel deexcitation pathway that consists of the 214- and 487-keV transitions. The γ rays at 421 and 463 keV are proposed to be the $4_2^+ \rightarrow 2_2^+$ and $5_1^+ \rightarrow 3_1^+$ transitions, respectively, based on the systematics of the quasi-γ-band levels for lighter Mo isotopes. Furthermore, the measured intensity of the 532-keV γ ray depopulating the yrast 6^+ state at 1132 keV implies a sizable value of spin ($\approx 5\hbar$) for the β-decaying state in ^{110}Nb, consistent with the argument above on the population of the 4_2^+ and 5_1^+ levels at 916 and 1164 keV, respectively, in the β decay of ^{110}Nb.

The experimental excitation energies of the 2_1^+, 4_1^+, and 2_2^+ states in even Mo isotopes are reproduced well in a microscopic theory based on the general Bohr Hamiltonian approach,[8] in which the potential energy and inertial functions (mass parameters) are calculated using the constrained Hartree-Fock-Bogoliubov (CHFB) method with the Skyrme effective interaction. For ^{110}Mo, the 2_2^+ state is predicted to lie at about 500 keV (see Fig. 6 in Ref.[8]), being in good agreement with the energy assigned in the present work. The level energies of the ground-state and quasi-γ bands calculated using the SIII and SLy4 versions of the Skyrme interaction are shown in Figs. 2(A) and 2(B), respectively. We also performed a similar calculation with the pairing-plus-quadrupole (P+Q) model including a quadrupole pairing,[9] in which the inertial functions are calculated with the local quasiparticle random-phase approximation (LQRPA). The result of this calculation is shown in Fig. 2(C).

Figure 2 provides the comparison of the experimental levels with the predictions of the two theoretical frameworks. The agreement between the observed and calculated level energies is very satisfactory for both the ground-state and quasi-γ bands. In particular, it is noteworthy that the observation for the 2_2^+ state being lower than the 4_1^+ level is reproduced in all of the calculations. In the right panels in Fig. 2, the potential energy surface calculations for ^{110}Mo exhibit local minima at the prolate ($\gamma \approx 0°$) and oblate ($\gamma \approx 60°$) sides using the SIII and SLy4 versions of the Skyrme interaction, respectively. However, the exact location of the energy minimum is

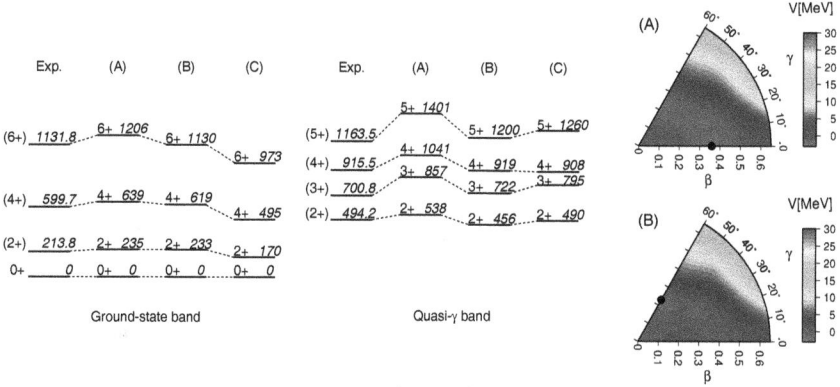

Fig. 2. Experimental and calculated level energies (in keV) for the ground-state (left) and quasi-γ bands (middle) in ^{110}Mo. The model calculations are based on the general Bohr Hamiltonian approach with the CHFB+IB method using the SIII (A) and SLy4 (B) versions of the Skyrme interaction, and the CHFB+LQRPA method using the P+Q force (C). Right panels show potential energy surface calculations for ^{110}Mo with the SIII (A) and SLy4 (B) versions of the Skyrme interaction. The energy minima are indicated with filled circles.

not important in characterizing the collective level properties of the heavier Mo isotopes, because an overall profile in the potential energy surface spreads over the γ degree of freedom. Consequently, the level structure of ^{110}Mo is ascribed to its γ-soft nature rather than the rigid deformation of any kind. More discussion on the structural evolution in even-even $A \approx 110$ nuclei is developed in Ref.[4] in comparison with the level properties in the neutron-rich $A \approx 190$ region.

References

1. Y. Yano, Nucl. Instrum. Methods B **261**, 1009 (2007)
2. T. Sumikama et al., Phys. Rev. Lett. **106**, 202501 (2011)
3. H. Watanabe et al., Phys. Lett. B **696**, 186 (2011)
4. H. Watanabe et al., Phys. Lett. B **704**, 270 (2011)
5. T. Kubo, Nucl. Instrum. Methods B **204**, 97 (2003)
6. T. Ohnishi et al., J. Phys. Soc. Jpn. **77**, 083201 (2008)
7. W. Urban et al., Eur. Phys. J. A **20**, 381 (2004)
8. L. Próchniak, Int. J. Mod. Phys. E **19**, 705 (2010)
9. N. Hinohara et al., Phys. Rev. C **82**, 064313 (2010)

New hidden beauty molecules predicted by the local hidden gauge approach and heavy quark spin symmetry

C. W. Xiao[1],[*], A. Ozpineci[2] and E. Oset[1]

[1] *Departamento de Física Teórica and IFIC, Centro Mixto Universidad de Valencia-CSIC, Institutos de Investigación de Paterna, Apartado 22085, 46071 Valencia, Spain*
[2] *Physics Department, Middle East Technical University, 06531 Ankara, Turkey*
[*] *E-mail: xiaochw@ific.uv.es*

Using a coupled channel unitary approach, combining the heavy quark spin symmetry and the dynamics of the local hidden gauge, we investigate the meson-meson interaction with hidden beauty. We obtain several new states of isospin $I = 0$: six bound states, and weakly bound six more possible states which depend on the influence of the coupled channel effects.

1. Introduction

The world of heavy quarks, charm and beauty, is experiencing a fast development, with the finding of new states in experiments. Recently, the discovery of the hidden beauty $Z_b(10610)$ and $Z_b(10650)$ states,[1] takes more attention to the beauty sector.[2,3] In this work, we investigate the hidden beauty system of meson-meson interaction.[4-6] We take into account the heavy quark spin symmetry (HQSS)[7-10] for the hidden beauty sector.

2. Formalism

In our work, we use the coupled channel approach to study the meson-meson interaction in the hidden beauty sector, which are the channels of $B^{(*)}_{(s)}\bar{B}^{(*)}_{(s)}$. Thus, in our case, all the hidden beauty systems are made by a meson(M) − antimeson($\bar{M}$) state. Under the HQSS constrain, we use the local hidden gauge approach[11,12] to evaluate the interaction potentials (more details, seen in our paper[13]), following development of Refs. 14,15.

3. Results

We use the coupled channel Bethe-Salpeter equation to evaluate the scattering amplitudes. For the G function, we take

$$G(s) = \int \frac{d^3\vec{q}}{(2\pi)^3} \, f^2(\vec{q}) \, \frac{\omega_1 + \omega_2}{2\,\omega_1\,\omega_2} \, \frac{1}{P^{0\,2} - (\omega_1 + \omega_2)^2 + i\varepsilon}, \quad f(\vec{q}) = \frac{m_V^2}{\vec{q}^2 + m_V^2},$$

where $f(\vec{q})$ is the form factor, which comes from the vector meson exchange.

Our results of the poles and couplings for the $J^{PC} = 2^{++}$ channel with $q_{max} = 415$ MeV (left panel) and $q_{max} = 830$ MeV (right panel), are shown as Table 1. When ignoring the coupled channel effect, the results are shown in Table 2. For the $J = 1$, $I = 0$ sector, the results with coupled channels and without coupled channels are shown in Tables 3 and 4. Finally, we get results for the $J^{PC} = 0^{++}$ sector as listing in Tables 5 and 6.

Table 1. The poles and couplings for the $J^{PC} = 2^{++}$: $q_{max} = 415$ MeV (left panel) and $q_{max} = 830$ MeV (right panel), all units in MeV.

10613	$B^*\bar{B}^*$	$B_s^*\bar{B}_s^*$	10469	$B^*\bar{B}^*$	$B_s^*\bar{B}_s^*$
g_i	86168	45864	g_i	174393	92843

Table 2. The poles and couplings for the $J^{PC} = 2^{++}$ ignoring coupled channels(two panels and units the same as before, also the same for below).

10616	$B^*\bar{B}^*$	$B_s^*\bar{B}_s^*$	10500	$B^*\bar{B}^*$	$B_s^*\bar{B}_s^*$
g_i	81595	0	g_i	159102	0
10828	$B^*\bar{B}^*$	$B_s^*\bar{B}_s^*$	10812	$B^*\bar{B}^*$	$B_s^*\bar{B}_s^*$
g_i	0	19787	g_i	0	44102

Table 3. The poles and couplings for the $J^{PC} = 1^{+-}$ and $J^{PC} = 1^{++}$.

10568	$B\bar{B}^*\pm$c.c.	$B_s\bar{B}_s^*\pm$c.c.	10425	$B\bar{B}^*\pm$c.c.	$B_s\bar{B}_s^*\pm$c.c.
g_i	85433	45560	g_i	172908	92232

4. Discussions

For a resonance or bound state, the sum rule is fulfilled: $P_p = -\sum_i g_i^2 \left[\frac{dG_i}{dE}\right]_{E=E_p} = 1$. For $B\bar{B}$ state, taking $q_{max} = 415$MeV, we get $P_{B\bar{B}} = 0.985$, which means that the bound state is mostly made by $B\bar{B}$ with a minor $B_s\bar{B}_s$ component. This $B\bar{B}$ state is stable and independent of the free parameters of our formalism, which can be seen in Table 7.

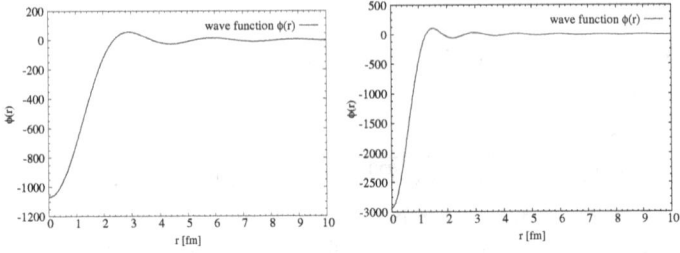

Fig. 1.   The wave functions of $B\bar{B}$ state, Left: $q_{max} = 415$MeV; Right: $q_{max} = 830$MeV.

Table 4.   The poles and couplings for the $J^{PC} = 1^{+-}$ and $J^{PC} = 1^{++}$ coupled channels. ignoring

10571	$B\bar{B}^*\pm$c.c.	$B_s\bar{B}_s^*\pm$c.c.	10455	$B\bar{B}^*\pm$c.c.	$B_s\bar{B}_s^*\pm$c.c.
g_i	80884	0	g_i	157691	0
10783	$B\bar{B}^*\pm$c.c.	$B_s\bar{B}_s^*\pm$c.c.	10768	$B\bar{B}^*\pm$c.c.	$B_s\bar{B}_s^*\pm$c.c.
g_i	0	19611	g_i	0	43776

Table 5.   The poles and couplings for the $J^{PC} = 0^{++}$.

10523	$B\bar{B}.$	$B_s\bar{B}_s$	10380	$B\bar{B}$	$B_s\bar{B}_s$
g_i	85045	45257	g_i	172046	91591

We also investigate the wave function and radius of the state. For the $B\bar{B}$ state, we show the results in Fig. 1. The radii of these states are given in Table 8, which are of the same order of magnitude as Refs. 2,16.

5. Conclusions

In our work, combining the local hidden gauge symmetry with heavy quark spin symmetry, we investigate the hidden beauty sector: $B^{(*)}_{(s)}\bar{B}^{(*)}_{(s)}$. In the $I = 0$ sector, we obtain 6 hidden beauty resonances with binding energies 34 MeV (178 MeV), and 6 hidden beauty-hidden strange states with binding energies 2 MeV (18 MeV). But, for the $I = 1$ sector, the interaction is too weak to form any bound states. We hope that these states can be found in the experiment in the future.

References

1. A. Bondar *et al.* [Belle Collaboration], Phys. Rev. Lett. **108**, 122001 (2012).
2. Z. -F. Sun, J. He, X. Liu, Z. -G. Luo and S. -L. Zhu, Phys. Rev. D **84**, 054002 (2011).

Table 6. The poles and couplings for the $J^{PC} = 0^{++}$ ignoring coupled channels.

10526	$B\bar{B}$.	$B_s\bar{B}_s$	10410	$B\bar{B}$	$B_s\bar{B}_s$
g_i	80528	0	g_i	156968	0
10738	$B\bar{B}$	$B_s\bar{B}_s$	10723	$B\bar{B}$	$B_s\bar{B}_s$
g_i	0	19441	g_i	0	43443

Table 7. The poles in the $J^{PC} = 0^{++}$ channel when the cut off is changed.

q_{max} (MeV)	450	500	600	700	800
pole (MeV)	10513	10498	10464	10427	10389

Table 8. The radii of the states.

states	$q_{max} = 415$MeV	$q_{max} = 830$MeV
$B^*\bar{B}^*$	1.46 fm	0.72 fm
$B\bar{B}^*$	1.46 fm	0.72 fm
$B\bar{B}$	1.46 fm	0.72 fm

3. M. Cleven, Q. Wang, F. -K. Guo, C. Hanhart, U. -G. Meissner and Q. Zhao, Phys. Rev. D 87, **074006** (2013).

4. Y. -J. Zhang, H. -C. Chiang, P. -N. Shen and B. -S. Zou, Phys. Rev. D **74**, 014013 (2006).

5. S. Ohkoda, Y. Yamaguchi, S. Yasui, K. Sudoh and A. Hosaka, Phys. Rev. D **86**, 034019 (2012).

6. M. T. Li, W. L. Wang, Y. B. Dong and Z. Y. Zhang, J. Phys. G **40**, 015003 (2013).

7. N. Isgur and M. B. Wise, Phys. Lett. B **232**, 113 (1989).

8. M. Neubert, Phys. Rept. **245**, 259 (1994).

9. F. -K. Guo, C. Hanhart and U. -G. Meissner, Phys. Rev. Lett. **102**, 242004 (2009).

10. C. Garcia-Recio, V. K. Magas, T. Mizutani, J. Nieves, A. Ramos, L. L. Salcedo and L. Tolos, Phys. Rev. D **79**, 054004 (2009).

11. M. Bando, T. Kugo, S. Uehara, K. Yamawaki and T. Yanagida, Phys. Rev. Lett. **54**, 1215 (1985).

12. M. Bando, T. Kugo and K. Yamawaki, Phys. Rept. **164**, 217 (1988).

13. A. Ozpineci, C. W. Xiao and E. Oset, Phys. Rev. D **88**, 034018 (2013).

14. J. -J. Wu, R. Molina, E. Oset and B. S. Zou, Phys. Rev. Lett. **105**, 232001 (2010).

15. J. -J. Wu, R. Molina, E. Oset and B. S. Zou, Phys. Rev. C **84**, 015202 (2011).

16. M. T. Li, W. L. Wang, Y. B. Dong and Z. Y. Zhang, Int. J. Mod. Phys. A **27**, 1250161 (2012).

Understanding the negative parity Λ resonances from the $K^-p \to \Lambda\eta$ reaction and their strong decays

Li-Ye Xiao and Xian-Hui Zhong[*]

Department of Physics, Hunan Normal University, and Key Laboratory of Low-Dimensional Quantum Structures and Quantum Control of Ministry of Education Changsha, 410081, China
[*] E-mail: zhongxh@hunnu.edu.cn

To understand the nature of the low-lying negative parity Λ resonances, we carried out a combined study of the reaction $K^-p \to \Lambda\eta$ at low energies and their strong decays with a chiral quark model. It is found that the low-lying negative parity Λ resonances, such as $\Lambda(1670)$, $\Lambda(1405)$, $\Lambda(1520)$ and $\Lambda(1690)$, are most likely mixed states between different configurations.

Keywords: Chiral quark model; resonances; strong decay.

1. Introduction

Our knowledge about Λ resonances is poor. Even for the well-established low-lying negative parity resonances, their properties are still controversial.[1] Recently, we systematically studied the reactions $K^-p \to \Sigma^0\pi^0, \Lambda\pi^0, \bar{K}^0 n$ in a chiral quark model.[2] We found that $\Lambda(1670)$ should have a very weak coupling to $\bar{K}N$, while $\Lambda(1520)$ needs a strong coupling to $\bar{K}N$, which cannot be well explained with the symmetry constituent quark model in the SU(6)⊗O(3) limit.[2]

To obtain more strong coupling properties and better understandings of these low-lying Λ states, we continue study another important reaction $K^-p \to \Lambda\eta$. Owing to isospin selection rule, only Λ resonances contribute here. Especially, the poorly known strong coupling of $\Lambda(1670)$ to $\Lambda\eta$ might be reliably obtained, for this reaction at threshold is dominated by formation of the $\Lambda(1670)$.[3]

Furthermore, to understand the natures of the strong coupling properties extracted from $\bar{K}N$ scattering, we further carry out a systematical study of the strong decays for the low-lying negative parity Λ resonances.

2. Results and Analysis

We can reasonably describe the experimental data of the $K^-p \to \Lambda\eta$ within our chiral quark model.[4] The total cross section of $K^-p \to \Lambda\eta$ compared with the data is shown in figure 1. It is found that $\Lambda(1670)S_{01}$ play a dominant role in the reactions. No obvious contribution from the D waves, $\Lambda(1520)$ and $\Lambda(1690)$, are seen, while u and t channel provide a large background. It should be pointed out that to reproduce the experimental data, $\Lambda(1670)S_{01}$ should have a much stronger coupling to $\Lambda\eta$ channel than that derived in the $SU(6) \otimes O(3)$ limit.

To further understand the properties of the low-lying negative parity Λ resonances extracted from the reactions, we study their strong decays in the chiral quark model as well. Considering configuration mixing effects in the low-lying negative Λ resonances and adopting the mixing schemes,

$$\begin{pmatrix} |\Lambda(1800)\frac{1}{2}^-\rangle \\ |\Lambda(1670)\frac{1}{2}^-\rangle \\ |\Lambda(1405)\frac{1}{2}^-\rangle \end{pmatrix} = \begin{pmatrix} 0.17 & 0.62 & 0.77 \\ 0.39 & -0.76 & 0.53 \\ 0.90 & 0.21 & -0.37 \end{pmatrix} \begin{pmatrix} |70,^2 1\rangle_S \\ |70,^2 8\rangle_S \\ |70,^4 8\rangle_S \end{pmatrix}, \qquad (1)$$

for the S-wave states, and

$$\begin{pmatrix} |\Lambda(1520)\frac{3}{2}^-\rangle \\ |\Lambda(1690)\frac{3}{2}^-\rangle \\ |\Lambda\frac{3}{2}^-\rangle_3 \end{pmatrix} = \begin{pmatrix} 0.94 & 0.34 & 0.09 \\ 0.31 & -0.92 & 0.26 \\ 0.17 & -0.21 & -0.96 \end{pmatrix} \begin{pmatrix} |70,^2 1\rangle_D \\ |70,^2 8\rangle_D \\ |70,^4 8\rangle_D \end{pmatrix}. \qquad (2)$$

for the D-wave states, we can well describe all the strong decay properties for the S- and D-wave states from the Particle Data Group.[5] Furthermore, with these mixing schemes, the $\Lambda\eta$ branching ratio for $\Lambda(1670)S_{01}$ is enhanced obviously, while the $N\bar{K}$ branching ratio is suppressed, which can naturally explain the weak coupling of $\Lambda(1670)S_{01}$ to $N\bar{K}$ and strong coupling to $\Lambda\eta$ needed in the $\bar{K}N$ reactions.

3. Summary

In this work, we have studied the low energy reaction $K^-p \to \Lambda\eta$ with a chiral quark model approach. A reasonable description of the measurements has been achieved. It is found that $\Lambda(1670)S_{01}$ dominates the reaction around at the low energy regions, and the non-resonant backgrounds also play crucial roles. The reaction indicates that $\Lambda(1670)S_{01}$ should have a much stronger coupling to $\Lambda\eta$ channel than that derived in the $SU(6) \otimes O(3)$ limit.

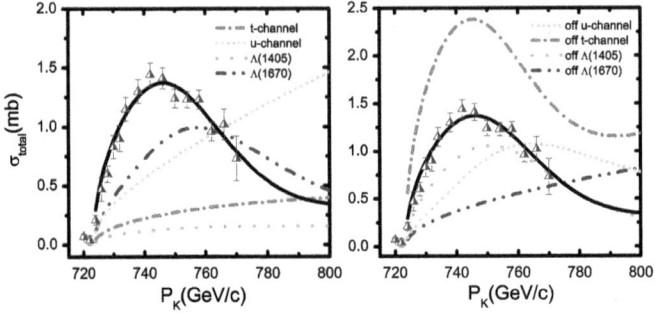

Fig. 1. The total cross section of $K^-p \to \Lambda\eta$ compared with the data from Ref. 3. The bold solid curves are for the full model calculations. In the left side, exclusive cross sections for $\Lambda(1405)S_{01}$, $\Lambda(1670)S_{01}$, $t-$ channel and $u-$channel are indicated explicitly by the legends in the figures. In the right side, the results by switching off the contributions of $\Lambda(1405)S_{01}$, $\Lambda(1670)S_{01}$, $t-$ channel and $u-$channel are indicated explicitly by the legends in the figures.

To understand these strong interaction properties of $\Lambda(1670)S_{01}$, we further study the strong decay properties of the low-lying negative parity Λ-resonances. It is found that the configuration mixing effects are crucial to understand the strong decay properties of the low-lying negative Λ resonances. These resonances are most likely mixed states between different configurations. Considering configuration mixing effects, we can reasonably explain the strong interaction properties of $\Lambda(1670)S_{01}$ extracted from the $K^-p \to \Lambda\eta$ reaction.

Acknowledgments

This work is supported, in part, by the National Natural Science Foundation of China (Grants No. 11075051 and No. 11375061), the Program Excellent Talent Hunan Normal University, the Hunan Provincial Natural Science Foundation (Grants No. 13JJ1018), and the Hunan Provincial Innovation Foundation For Postgraduate.

References

1. E. Klempt and J. -M. Richard, Rev. Mod. Phys. **82**, 1095 (2010).
2. X. -H. Zhong and Q. Zhao, Phys. Rev. C **88**, 015208 (2013).
3. A. Starostin *et al.* [Crystal Ball Collaboration], Phys. Rev. C **64**, 055205 (2001).

4. L. -Y. Xiao and X. -H. Zhong, Phys. Rev. C **88** (2013) 065201 [arXiv:1309.1923 [nucl-th]].
5. J. Beringer *et al.* [Particle Data Group], Phys. Rev. D **86**, 010001 (2012).

Exotic dibaryons with a heavy antiquark

Y. Yamaguchi* and A. Hosaka

*Research Center for Nuclear Physics (RCNP), Osaka University
Ibaraki, Osaka, 567-0047, Japan
* E-mail: yamaguti@rcnp.osaka-u.ac.jp*

S. Yasui

*KEK Theory Center, Institute of Particle and Nuclear Studies, High Energy
Accelerator Research Organization
1-1, Oho, Ibaraki, 305-0801, Japan*

The possible existence of $\bar{D}NN$ and BNN states is discussed. They are manifestly exotic dibaryons whose bound states which are stable against a strong decay. As for the $\bar{D}^{(*)}N$ ($B^{(*)}N$) interactions, we consider the one pion exchange potential enhanced by the heavy quark spin symmetry. By solving the coupled-channel Schrödinger equations for the three-body systems, we find the bound states with $J^P = 0^-$ and resonances with $J^P = 1^-$ for $I = 0$. We also discuss the spin degeneracy of the $P_Q NN$ states in the heavy quark mass limit.

Keywords: Exotic dibaryons; Heavy quark spin symmetry; Heavy mesons; one pion exchange potential.

1. Introduction

The hadron-nucleon interaction gives us rich phenomena which are not seen in normal nuclei. In the strangeness sector, the hadrons bound in nuclei cause the impurity effects such as the formation of the high density states in the $\bar{K}$ nuclei,[1] and the shrinking of the wave functions in the hypernuclei.[2]

Recently, for the heavy flavor sector, the sufficiently strong attraction between a $\bar{D}$ (B) meson and a nucleon N have been discussed.[3–6] The attraction is produced by the one pion exchange potential (OPEP) with the heavy quark spin symmetry (HQS), and it leads to two-body $\bar{D}N$ (BN) molecules. They are manifestly exotic states having no lower hadronic channels coupled by a strong interaction.

The attraction in the $\bar{D}N$ (BN) forces motivates us to explore the exotic nuclear systems with a heavy quark. In fact, there have been many works for $\bar{D}$ (B) mesons in nuclear systems with large baryon numbers (resent results are summarized in Ref. 7). However, $\bar{D}$ (B) nuclei as few-body systems have not been investigated so far in the literature. The few-body systems would be more likely to be produced in the hadron colliders.

In the present work, we study the mass spectrum of $\bar{D}NN$ and BNN bound and/or resonant states. We also investigate $P_Q NN$ states, where P_Q is defined as a heavy meson having an infinite heavy quark mass.

2. Interactions

Let us discuss the basic interaction for the $P^{(*)}N$ ($P^{(*)} = \bar{D}^{(*)}, B^{(*)}$). In the $P^{(*)}N$ systems, the OPEP enhanced by the HQS is the basic ingredient to provide a strong attraction. The HQS manifests the mass degeneracy of heavy pseudoscalar meson P and heavy vector meson P^*.[8] Indeed, the mass splitting between P and P^* mesons is small, $m_{\bar{D}^*} - m_{\bar{D}} \sim 140$ MeV and $m_{B^*} - m_B \sim 45$ MeV. Thanks to the mass degeneracy, the OPEP causing the couplings of $PN - P^*N$ and $P^*N - P^*N$ are enhanced.

The PN force could also have the short-range parts. However, we expect that the OPEP as a long-range force dominates when the systems form a loosely bound state. An analogous situation occurs for the deuteron.

Let us show the OPEP between $P^{(*)}$ and N briefly. Details are given in Ref. 4–6. The OPEPs for $PN - P^*N$ and $P^*N - P^*N$ are given by

$$V_{PN-P^*N}(r) = -\frac{g_\pi g_{\pi NN}}{\sqrt{2}m_N f_\pi} \frac{1}{3} \left[\vec{\varepsilon}^{\,\dagger} \cdot \vec{\sigma} C(r) + S_\varepsilon T(r) \right] \vec{\tau}_P \cdot \vec{\tau}_N, \qquad (1)$$

$$V_{P^*N-P^*N}(r) = \frac{g_\pi g_{\pi NN}}{\sqrt{2}m_N f_\pi} \frac{1}{3} \left[\vec{S} \cdot \vec{\sigma} C(r) + S_S T(r) \right] \vec{\tau}_P \cdot \vec{\tau}_N, \qquad (2)$$

as a sum of the central and tensor forces, $C(r)$ and $T(r)$, in Ref. 4–6. The coupling constant for $P^{(*)}P^*\pi$ ($NN\pi$) vertex is $g_\pi = 0.59$ ($g_{\pi NN}^2/4\pi = 13.6$). We note the $PN - PN$ term is absent due to parity conservation.

As for the NN interaction, we employ the Argonne v_8' potential,[9] which is formed by the central forces with operators $[1, (\vec{\sigma}_1 \cdot \vec{\sigma}_2)] \otimes [1, (\vec{\tau}_1 \cdot \vec{\tau}_2)]$, the tensor forces with $S_{12} \otimes [1, (\vec{\tau}_1 \cdot \vec{\tau}_2)]$, and the LS forces with $(\vec{L} \cdot \vec{S}) \otimes [1, (\vec{\tau}_1 \cdot \vec{\tau}_2)]$.

The Hamiltonian is given by $H = T + V_{P^{(*)}N} + V_{NN}$, where T is the kinetic term, and $V_{P^{(*)}N}$ (V_{NN}) is the $P^{(*)}N$ (NN) potential shown above. By diagonalizing the Hamiltonian, we obtain the eigenenergies. The three-body wave functions are expressed by the Gaussian expansion method.[10] The poles of resonances are calculated by the complex scaling method.[11]

3. Numerical Results

In this section, the results for the bound and resonant states in $\bar{D}NN$, BNN and $P_Q NN$ are shown. The obtained energies are summarized in Table 1. The binding energy is given as a real negative value, and the resonance energy E_{re} and decay width Γ are given as $E_{\mathrm{re}} - i\Gamma/2$. The energies are measured from the lowest thresholds.

First, let us present the results of $\bar{D}NN$ and BNN for $J^P = 0^-$. We obtain bound states with $(I, J^P) = (1/2, 0^-)$. The binging energies are -5.2 MeV for $\bar{D}NN$ and -26.2 MeV for BNN. We find that the BNN state is more bound than the $\bar{D}NN$ state, because the $PNN - P^*NN$ mixing effects are enhanced, when P and P^* mesons become more degenerate.

We analyze the expectation values of the potentials, V_{PN-P^*N}, $V_{P^*N-P^*N}$ and V_{NN}, summarized in Table 2. We find that the tensor forces of V_{PN-P^*N} provide the dominant contribution both in $\bar{D}NN$ and BNN. For V_{NN}, we find the tensor force in V_{NN} which is a driving force in the deuteron d is almost irrelevant, while the central force is rather dominant. This is reasonable because d (with quantum number 1^+) does not exist in the main component of $\bar{D}NN$ with $J^P = 0^-$.

In scattering states, we find resonances both for $\bar{D}NN$ and BNN with $(I, J^P) = (1/2, 1^-)$. We obtain $E_{\mathrm{re}} - \Gamma/2 = 111.2 - i9.3$ MeV for $\bar{D}NN$, and $6.8 - i0.2$ MeV for BNN, respectively. When we consider only the P^*NN channel, we obtain a bound state of P^*NN. The results indicate that these states are Feshbach resonances.

Finally, we consider $P_Q NN$ systems with $m_{P_Q^*} - m_{P_Q^*} = 0$. Interestingly, we find the degenerate bound states for $J^P = 0^-$ and 1^- with the same binging energy -38.5 MeV. We find that the spin degeneracy due to the HQS, discussed in Refs. 8,12, is realized in the $P_Q NN$ states.

Table 1. Energies of $\bar{D}NN$, BNN and $P_Q NN$ from Ref. 6. All values are given in units of MeV.

(I, J^P)	$\bar{D}^{(*)}NN$	$B^{(*)}NN$	$P_Q^{(*)}NN$
$(1/2, 0^-)$	-5.2	-26.2	-38.5
$(1/2, 1^-)$	$111.2 - i9.3$	$6.8 - i0.2$	-38.5

Table 2. Expectation values of central, tensor and LS forces of the $\bar{D}^{(*)}N$ $(B^{(*)}N)$ and NN potentials in the bound state of $\bar{D}NN$ (BNN). All values are in units of MeV.

PNN	$\langle V_{\bar{D}N-\bar{D}^*N}\rangle$	$\langle V_{\bar{D}^*N-\bar{D}^*N}\rangle$	$\langle V_{NN}\rangle$	$\langle V_{BN-B^*N}\rangle$	$\langle V_{B^*N-B^*N}\rangle$	$\langle V_{NN}\rangle$
Central	-2.3	-0.1	-9.5	-6.5	0.3	-11.6
Tensor	-47.1	0.7	-0.2	-92.0	-2.7	-1.0
LS	—	—	-0.03	—	—	-0.1

4. Summary

We have explored the possible existence of and $P_Q NN$. The OPEP and Argonne v'_8 potential were employed as the $P^{(*)}N$ and NN interactions, respectively. By solving the coupled-channel equations for PNN and P^*NN, we have obtained bound states with $J^P = 0^-$ and Feshbach resonances with $J^P = 1^-$ for $I = 1/2$ both in $\bar{D}NN$ and BNN. The tensor force of the OPEP mixing PN and P^*N plays an important role to produce a strong attraction. For the $P_Q NN$ systems we have obtained degenerate bound states of $J^P = 0^-$ and 1^- for $I = 0$.

The $\bar{D}NN$ and BNN states can be searched in relativistic heavy ion collisions in RHIC and LHC.[13] Furthermore, the search for the $\bar{D}NN$ would be also carried out in J-PARC and GSI-FAIR.

Acknowledgments

The authors would like to thank Dr. Y. Kikuchi for valuable discussions and fruitful suggestions. This work is supported in part by Grant-in-Aid for "JSPS Fellows(24-3518)"(Y. Y.) from Japan Society for the Promotion of Science.

References

1. T. Yamazaki, A. Dote and Y. Akaishi, Phys. Lett. B **587** (2004) 167.
2. O. Hashimoto and H. Tamura, Prog. Part. Nucl. Phys. **57** (2006) 564.
3. T. D. Cohen, P. M. Hohler and R. F. Lebed, Phys. Rev. D **72** (2005) 074010.
4. S. Yasui and K. Sudoh, Phys. Rev. D **80** (2009) 034008.
5. Y. Yamaguchi, S. Ohkoda, S. Yasui and A. Hosaka, Phys. Rev. D **84** (2011) 014032; Phys. Rev. D **85** (2012) 054003.
6. Y. Yamaguchi, S. Yasui and A. Hosaka, arXiv:1309.4324 [nucl-th].
7. S. Yasui and K. Sudoh, Phys. Rev. C **87** 015202 (2013).
8. A. V. Manohar and M. B. Wise, Camb. Monogr. Part. Phys. Nucl. Phys. Cosmol. **10**, 1-191 (2000).
9. B. S. Pudliner, V. R. Pandharipande, J. Carlson, S. C. Pieper and R. B. Wiringa, Phys. Rev. C **56** (1997) 1720.
10. E. Hiyama, Y. Kino and M. Kamimura, Prog. Part. Nucl. Phys. **51** (2003) 223.
11. S. Aoyama, T. Myo, K. Katō and K. Ikeda, Prog. Theor. Phys. **116** (2006) 1
12. S. Yasui, K. Sudoh, Y. Yamaguchi, S. Ohkoda, A. Hosaka and T. Hyodo, Phys. Lett. B **727** (2013) 185.
13. S. Cho *et al.* [ExHIC Collaboration], Phys. Rev. C **84** (2011) 064910.

Recent progress towards a chiral effective field theory for the NN system

C. J. Yang[*] and Bingwei Long[†]

Dipartimento di Fisica, Universita di Trento, via Sommarive, 14 I-38123 Trento, Italy
Department of Physics, Sichuan University, 29 Wang-Jiang Road, Chengdu, Sichuan 610064, China
[*] *E-mail: chieh@science.unitn.it*
[†] *E-mail: bingwei@scu.edu.cn*

Since Weinberg's proposal two decades ago, chiral effective field theory in the NN sector has been developed and applied up to order $O((Q/M_{hi})^4)$. In principle it could provide a model-independent description of nuclear force from QCD. However, in spite of its huge success, some open issues such as the renormalization group invariance and power counting, still remain to be solved. In this talk we refine the chiral effective field theory approach to the NN system based on a renormalization group analysis. Our results show that a truly model-independent description of NN system can be obtained by a new power counting which treats the subleading order corrections perturbatively.

1. Introduction

Based on the symmetries of Quantum chromodynamics (QCD) in the low energy region (≤ 1 GeV), chiral effective field theory (χEFT) enables calculations of strong interaction in the non-perturbative region. However, unlike the pion-pion and pion-nulceon section, where the power counting—the key ingredient which guarantees the intrinsic consistency of an EFT—is given clearly from the vertices generated by the chiral Lagrangian, the power counting in nucleon-nucleon (NN) case is hindered by the infrared enhancement and cannot be obtained straightforwardly.

The first step out of the NN problem, as suggested by Weinberg,[1] is to apply the power counting to the NN potential level first, and then sum the amplitude by iterating the potential in Schrodinger or Lippmann-Schwinger (LS) equation with an ultraviolet cutoff Λ. Currently, this prescription (the so-call Weinberg power counting (WPC)) has been carried out to next-to-next-to-next-to leading order (N³LO)[a], and has became the standard of many conventional calculations.[2–4] However, since Weinberg prescription only applies power counting to the potential level, the systematic control of the theory could be lost in the final amplitude.

2. Renormalization Group Analysis

One way to check whether a proposed scheme is under control is to perform the RG-analysis. RG-analysis carried out at leading order (LO), up to next-to-next-to leading order (NNLO) and N³LO based on WPC indicate that the conventional implementation of WPC fails to fulfill the RG requirement once the ultraviolet cutoff of the iteration $\Lambda > 1$ GeV.[5–8] Since the chiral expansion is established in powers of Q/M_{hi}[b], it is natural to question that the theory is no longer valid by bringing the intermediate state to $p \sim \Lambda > M_{hi}$. Thus, it makes little sense to perform RG-analysis for $\Lambda > 1$ GeV, even the final on-shell $Q << M_{hi}$.[9]

A second point of view[10–16] takes the final amplitude as a partial sum of the (infinitely many) diagrams, then under the assumption that a reasonable separation of scales exists[c], in any EFT one should be able to organize those diagrams in a systematic way to absorb the unimportant physics into contact term(s) order by order after a proper renormalization. Thus, as long as $Q << M_{hi}$, the impact of high-energy physics (which is well-represented by the contact term(s)) in the final amplitude should reduce as the increase of Λ, since the contribution from physics haven't been integrated out (i.e., from Λ to ∞) becomes smaller and smaller.

The answer of the above in-debating issue actually depends on how the diagrams are organized. It was shown that due to the fine-tuning of low energy constants and a Wigner bond-like effect,[17] once a cutoff $\Lambda > 1$ GeV is adopted the renormalization is effectively dominated by one contact term under the WPC scheme.[8] Moreover, a full-iteration of some type of irreducible two-pion-exchange diagrams could result in a pole-like structure.[18] Therefore, if one insists to build a NN potential based on χEFT and utilizes it later in a conventional way (e.g., inserts it as a potential in Schrodinger or LS equation), he or she needs to stay in $500 < \Lambda < 1000$ MeV. The consequence is that a full RG-based analysis becomes inapplicable.

---

[a] Note that the order here is defined based on the pion-exchange (long-range) part of the potential, which does not necessary equal to the order at the final NN amplitude.
[b] Here $Q \equiv (p, m_\pi)$, p the NN c.m. momentum, m_π the pion mass, and the breakdown scale M_{hi} is nominally $m_\rho \sim 4\pi f_\pi$.
[c] In the case where there is no reasonable separation of scales, EFT is impossible.

To allow a full RG-analysis, one must give up treating the whole chiral potential non-perturbatively. In other words, for the NN case there exists no "ideal potential" (in the traditional sense) to be extracted or derived. Some parts of the diagrams have to be included perturbatively. Recent works[10–13,15,16] which treat the subleading chiral potentials in the framework of Distorted-Wave-Born-Approximation enable a full RG- and power counting analysis. Once the $\Lambda-$dependence is under well-control, the estimation of the theoretical error becomes much easier, i.e., the error is given by $O(Q^{n+1}/M_{hi}^{n+1})$ up to order-n in the new power counting scheme.

We must point out that the lacks of a RG-analysis cannot rule out the possibility that WPC under the specified range of Λ could generate final amplitudes which has the correct power counting, but here is no way to check this so far. On the other hand, a RG-correct scheme could converge too slowly to be useful. Thus, before the full implementation to few- and manybody calculations, one cannot determine the superiority of either scheme. Nevertheless, it is of importance to start with a scheme which allows a full RG-analysis first, then check the power counting step by step to build the theory on a more solid ground.

3. New Power Counting and Future Task

The new power counting developed so far[11–13] can be summarized as:

1. The LO potential needs to be iterated to all order[d], and all subleading chiral potentials are included perturbatively as represented diagrammatically in Fig. 1.

2. The contact terms are determined by RG-analysis to guarantee the correct RG-behavior. As a general rule, for potentials which are singular and attractive at LO (i.e., $V_{LO}(r \rightarrow 0) \approx -\frac{1}{r^n}$ with $n \geq 3$), all contact terms need to be promoted one order earlier with respect to WPC.

Phase shifts from the above new scheme are evaluated up to NNLO, and the agreement with the Nijmegen phase-shift analysis[19] is comparable to those from WPC at the same order.

Furture tasks such as refining power counting with Lepage plot[20] or similar techniques, including Delta (1232) contribution, and deciding the power counting for high partial-waves ($l \geq 2$) are under investigation.

---

[d]At least for those spin-triplet and $l \leq 1$ partial-waves.

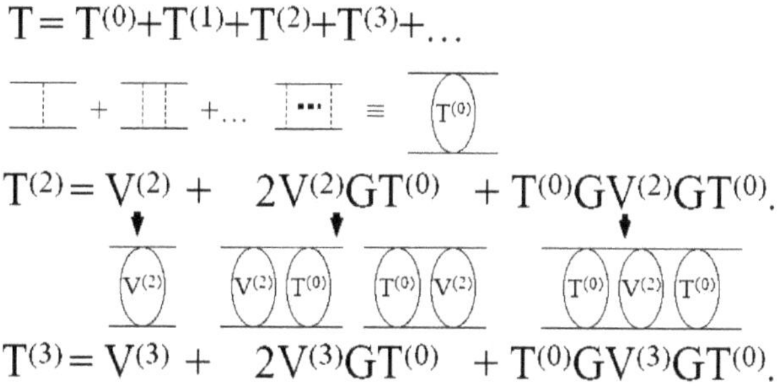

$$T = T^{(0)} + T^{(1)} + T^{(2)} + T^{(3)} + \ldots$$

$$T^{(2)} = V^{(2)} + 2V^{(2)}GT^{(0)} + T^{(0)}GV^{(2)}GT^{(0)}.$$

$$T^{(3)} = V^{(3)} + 2V^{(3)}GT^{(0)} + T^{(0)}GV^{(3)}GT^{(0)}.$$

Fig. 1.   Diagrammatic representation of the new power counting in the case where the $O(Q)$ contribution is absent. Here T and V denotes the T-matrix and chiral potential with their order indicate in the superscript.

Acknowledgments

This work is supported by the US NSF under grant PHYS-0854912, the INFN grant 40101979-40300059 and US DOE under contract No. DE-FG02-04ER41338, DE-AC05-06OR2317, E-AC05-06OR23177 and DE-FG02-93ER40756.

References

1. S. Weinberg, Phys. Lett. B **251**, 288 (1990); Nucl. Phys. B **363**, 3 (1991).
2. D. R. Entem and R. Machleidt, Phys. Rev. C **68**, 041001(R) (2003).
3. E. Epelbaum, W. Glöckle and U-G. Meißner, Nucl. Phys. A **747**, 362 (2005).
4. E. Epelbaum and U.-G. Meissner, Ann. Rev. Nucl. Part. Sci. **62** 159-185 (2012).
5. A. Nogga, R. Timmermans and U. van Kolck, Phys. Rev. C **72** 054006 (2005).
6. C.-J. Yang, Ch. Elster and D. R. Phillips, Phys. Rev. C **80** 034002 (2009).
7. C.-J. Yang, Ch. Elster and D. R. Phillips, Phys. Rev. C **80** 044002 (2009).
8. Ch. Zeoli, R. Machleidt and D. R. Entem, Few-body syst., **54**, 12, 2191-2205.
9. E. Epelbaum and G. Gegelia, Eur. Phys. J. A**41** 341-354 (2009).
10. M. C. Birse, PoS CD **09**, 078 (2009) [arXiv:0909.4641 [nucl-th]].
11. Bingwei Long and C.-J. Yang, Phys. Rev. C **84** 057001 (2011).
12. Bingwei Long and C.-J. Yang, Phys. Rev. C **86** 024001 (2012).
13. Bingwei Long and C.-J. Yang, Phys. Rev. C **85** 034002 (2012).
14. Bingwei Long, Phys. Rev. C **88** 014002 (2013).
15. M. P. Valderrama, Phys. Rev. C **83** 024003 (2011).
16. M. P. Valderrama, Phys. Rev. C **84** 064002 (2011).
17. E. P. Wigner, Phys. Rev. **98**, 145 (1955).
18. V. Baru *et al.*, Eur. Phys. J. A **48** 69 (2012).
19. V. G. J. Stoks *et al.*, Phys. Rev. C **48**, 792 (1993).
20. G. P. Lepage, arXiv:nucl-th/9706029.

Progress in resolving charge symmetry violation in nucleon structure

R. D. Young*, P. E. Shanahan and A. W. Thomas

ARC Centre of Excellence in Particle Physics at the Terascale and CSSM,
School of Chemistry and Physics, University of Adelaide,
Adelaide SA 5005, Australia
* E-mail: ross.young@adelaide.edu.au

Recent work unambiguously resolves the level of charge symmetry violation in moments of parton distributions using (2+1)-flavor lattice QCD. We introduce the methods used for that analysis by applying them to determine the strong contribution to the proton–neutron mass difference. We also summarize related work which reveals that the fraction of baryon spin which is carried by the quarks is in fact structure-dependent rather than universal across the baryon octet.

Keywords: Charge symmetry; lattice QCD; parton distributions; isospin.

1. Introduction

Charge symmetry, the equivalence of the u quark in the proton and the d in the neutron, and vice versa, is an excellent approximation in nuclear and hadronic systems — typically respected at $\sim 1\%$ precision.[1-3] Current deep inelastic scattering measurements are such that this level of precision has not yet been reached, with current bounds on charge symmetry violation (CSV) in parton distributions in the range 5-10%.[4] Such possibly large CSV effects are of particular interest in the context of a new program at Jefferson Laboratory[5] which aims to measure the electron-deuteron parity-violating deep inelastic scattering (PVDIS) asymmetry to better than 1% precision. This would offer an improvement of roughly an order of magnitude over early SLAC measurements,[6] with the potential to constitute an important new test of the Standard Model. Reaching this goal will rely on a precise control of strong interaction processes. CSV is likely to be the most significant hadronic uncertainty at the kinematics typical of the JLab program.[7-9] Phenomenological studies suggest that CSV could cause $\sim 1.5 - 2\%$ variations in the PVDIS asymmetry.[4] This is sufficient to disguise any signature of new physics, such as supersymmetry, expected to appear at the 1% level.[10]

Here we review our recent work[11] which has determined the CSV moments of parton distributions from lattice QCD. Our results, based on $(2 + 1)$-flavor lattice QCD simulations,[12,13] reveal $\sim 0.20 \pm 0.06\%$ CSV in the quark momentum fractions. This corresponds to a $\sim 0.4 - 0.6\%$ correction to the PVDIS asymmetry. This precision represents an order of magnitude improvement over the phenomenological bounds reported in Ref. 4. This result also constitutes an important step towards resolving the famous NuTeV anomaly.[14,15] Whereas the original report of a 3-sigma discrepancy with the Standard Model was based on the assumption of negligible CSV, effects of the magnitude and sign reported here act to reduce this discrepancy by one sigma. Similar results for spin-dependent parton CSV suggest corrections to the Bjorken sum rule[16] at the half-percent level which could possibly be seen at a future electron collider.[17]

In Sec. 2 we introduce the techniques used for our calculation in the context of the octet baryon mass splittings.[18] Section 3 summarizes our parton CSV results, presented in full in Ref. 11. Related work which reveals that the fraction of baryon spin which is carried by the quarks is in fact structure-dependent rather than universal across the baryon octet[19] is highlighted in Sec. 4.

2. Baryon Mass Splittings

Charge symmetry refers to the invariance of the strong interaction under a $180°$ rotation about the '2' axis in isospin space. At the parton level this invariance implies the equivalence of the u quark in the proton and the d quark in the neutron, and vice-versa. The symmetry would be exact if

- the up and down quarks were mass degenerate: $m_u = m_d$
- the quark electromagnetic charges were equal: $Q_u = Q_d$.

Of course, both of these conditions are broken in nature. This breaking manifests itself, for example, as mass differences between members of baryon isospin multiplets. While these differences have been measured extremely precisely experimentally,[20] the decomposition of these quantities into strong (from $m_u \neq m_d$) and electromagnetic (EM) contributions is much less well known. Phenomenological best estimates come from an application of the Cottingham sum rule[21] which relates the electromagnetic baryon self-energy to electron scattering observables. Walker-Loud, Carlson & Miller (WLCM) have recently revised the standard Cottingham formula;[22] noting that two

Lorentz equivalent decompositions of the $\gamma N \to \gamma N$ Compton amplitude produce inequivalent self-energies, WLCM use a subtracted dispersion relation to remove the ambiguity. This revision modifies traditional values of the EM part of the baryon mass splittings.

It is clearly valuable to independently determine either the strong or EM contribution to the proton–neutron mass difference. In principle this is achievable with lattice QCD. At this time, however, most lattice simulations for the octet baryon masses are performed with 2+1 quark flavours, that is, with mass-degenerate light quarks: $m_u = m_d$. Our analysis uses isospin-averaged lattice simulation results[23,24] to constrain chiral perturbation theory expressions for the baryon masses. Because of the symmetries of chiral perturbation theory, the only additional input required to determine the strong contribution to the baryon mass splittings is the up-down quark mass ratio m_u/m_d. The remainder of this section is devoted to an illustration of this method.

The usual meson-baryon Lagrangian can be written

$$\mathcal{L}^B = i \operatorname{Tr} \overline{\mathbf{B}}(v \cdot \mathcal{D})\mathbf{B} + 2D \operatorname{Tr} \overline{\mathbf{B}} S^\mu \{A_\mu, \mathbf{B}\} + 2F \operatorname{Tr} \overline{\mathbf{B}} S^\mu [A_\mu, \mathbf{B}]$$
$$+ 2b_D \operatorname{Tr} \overline{\mathbf{B}} \{\mathcal{M}_q, \mathbf{B}\} + 2b_F \operatorname{Tr} \overline{\mathbf{B}} [\mathcal{M}_q, \mathbf{B}]$$
$$+ 2\sigma_0 \operatorname{Tr} \mathcal{M}_q \operatorname{Tr} \overline{\mathbf{B}}\mathbf{B}.$$

The D and F terms denote the meson–baryon interactions and generate the nonanalytic quark mass dependence associated with quantum fluctuations of the pseudo-Goldstone modes. The explicit quark mass dependence is carried by the mass matrix $\mathcal{M}_q$, which is related to only three undetermined low-energy constants: b_D, b_F and σ_0 (at this order). With these constants determined by a fit to isospin-averaged (2+1-flavor) lattice data, there are no new parameters in the effective field theory relevant to CSV. Combined with appropriate treatment of the CSV loop corrections, our analysis of two independent lattice simulations yields the charge symmetry-breaking derivative[19]

$$m_\pi^2 \frac{d}{d\omega}(M_n - M_p) = (20.3 \pm 1.2)\,\mathrm{MeV} \quad [\text{PACS-CS}]$$

$$m_\pi^2 \frac{d}{d\omega}(M_n - M_p) = (16.6 \pm 1.2)\,\mathrm{MeV} \quad [\text{QCDSF}].$$

Here the quark mass splitting is denoted by ω, which is related to the quark mass ratio $(R = m_u/m_d)$ by

$$\omega = \frac{1}{2}\frac{(1-R)}{(1+R)} m_{\pi(\text{phys})}^2. \tag{1}$$

The dependence of our determination of $(M_p - M_n)^{\text{Strong}}$ on the input quark mass ratio is indicated in Fig. 1. In Fig. 2 this analysis, where we consider

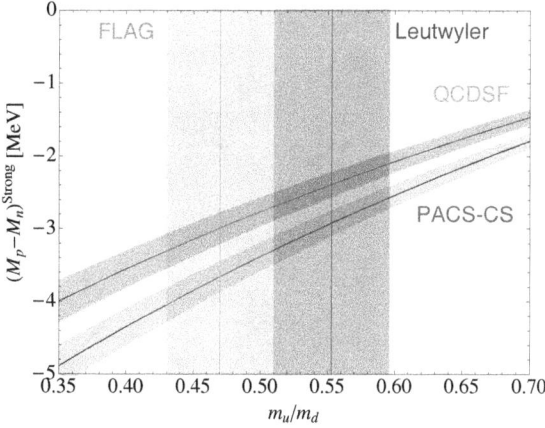

Fig. 1.   Strong nucleon mass splitting from our analysis of two independent lattice simulations (QCDSF[24] and PACS-CS[23]), plotted against the quark mass ratio m_u/m_d. Phenomenological (Leutwyler[25]) and lattice (FLAG[26]) values for this ratio are shown.

both PACS-CS and Leutwyler results and allow for both Leutwlyer and FLAG values of the ratio m_u/m_d, is compared against a recent strong mass splitting calculation of the BMW Collaboration[27] and the phenomenological estimates of the electromagnetic self energy.[21,22] Only for the purpose of simplifying the graphic have we not shown other recent lattice QCD estimates of the strong contribution to the mass splitting.[28–31]

3. CSV Parton Distribution Moments

The spin-independent CSV Mellin moments are defined as

$$
\begin{aligned}
\delta u^{m\pm} &= \int_0^1 dx\, x^m \left(u^{p\pm}(x) - d^{n\pm}(x) \right) \\
&\quad - \langle x^m \rangle_u^{p\pm} - \langle x^m \rangle_d^{n\pm},
\end{aligned}
$$

$$
\begin{aligned}
\delta d^{m\pm} &= \int_0^1 dx\, x^m \left(d^{p\pm}(x) - u^{n\pm}(x) \right) \\
&= \langle x^m \rangle_d^{p\pm} - \langle x^m \rangle_u^{n\pm},
\end{aligned}
$$

with similar expressions for the analogous spin-dependent terms $\delta\Delta q^{\pm}$. Here, the plus (minus) superscripts indicate C-even (C-odd) distributions $q^{\pm}(x) = q(x) \pm \bar{q}(x)$.

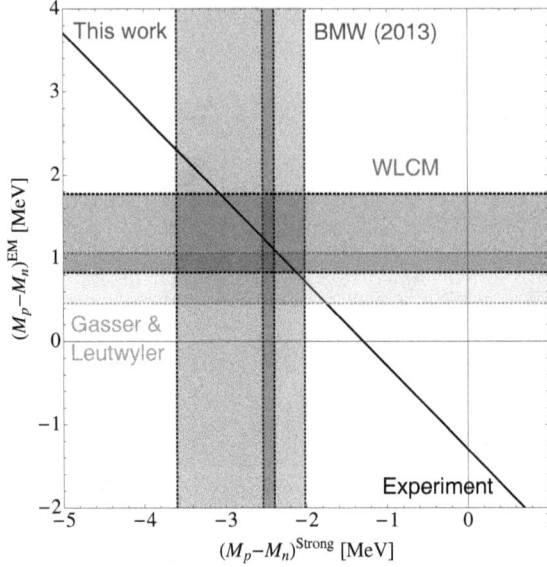

Fig. 2. Status of the nucleon mass splitting decomposition. Gasser-Leutwyler[21] and WLCM[22] calculations of the electromagnetic contribution are compared with the strong contribution determined in this work[19] and by the BMW lattice collaboration.[27] The black line indicates the experimental determination of the total mass difference.[20]

The first two spin-dependent and first spin-independent lattice-accessible moments have recently been determined from $(2 + 1)$-flavor lattice QCD by the QCDSF/UKQCD Collaboration.[12,13] These original papers made first estimates for the amount of CSV in the parton moments by considering the leading flavour expansion about the SU(3) symmetric point.[12,13] In Ref. 11 we applied an SU(3) chiral expansion in the same fashion as the baryon mass expansion described above. This enabled us to extrapolate the results away from the SU(3) symmetric point to determine the CSV contribution at the physical quark masses. Although this work only determines the lowest nontrival spin-independent moment, we can infer the CSV distribution as shown in Fig. 3 by using the same parameterisation of the x dependence as Ref. 4.

This magnitude of charge symmetry breaking is found to be in agreement with phenomenological MIT bag model estimates.[32,33] This result is of particular significance in the context of a new program to measure the (PVDIS) asymmetry to high precision at Jefferson Laboratory.[5,34] Further, the sign and magnitude of these results suggest a 1-σ reduction of the NuTeV anomaly.[15]

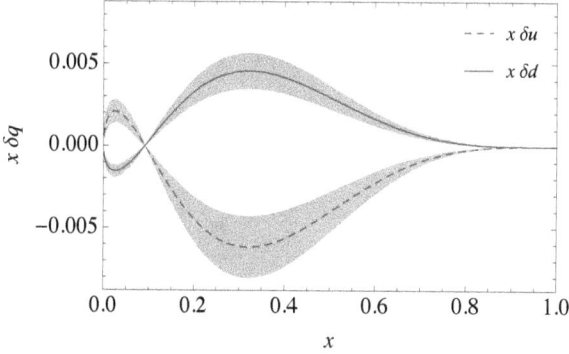

Fig. 3. Charge symmetry violating momentum fraction using simple phenomenological parameterisation $\delta q(x) = \kappa x^{-1/2}(1-x)^4(x-1/11)$ with normalisation determined from the lattice moment.[11]

4. Octet Baryon Spin Fractions

In addition to using the chiral extrapolation of the previous section to extract CSV effects, we have also determined the relative quark spin fractions in the octet baryons.[18] Figure 4, taken from Ref. 18, illustrates that the quark spin fraction is environment dependent. The figure clearly highlights that this result is evident in the bare lattice results, with considerable enhancement seen in the extrapolation to the physical point. Clearly, any candidate explanation of the proton spin problem must allow for the fraction of spin carried by the quarks to be dependent on baryon structure.

This finding is supported by a Cloudy Bag Model calculation, which includes relativistic and one-gluon-exchange corrections.[35-37] Within this model, the observed variation in quark spin arises from the meson cloud correction being considerably smaller in the Ξ than in the nucleon. That, combined with the less relativistic motion of the heavier strange quark, results in the total spin fraction in the Ξ being significantly larger than in the nucleon.

5. Conclusion

The effects of charge symmetry violation (CSV) are becoming increasingly significant in precision studies of the Standard Model. Recent results, based on $(2 + 1)$-flavor lattice QCD simulations, unambiguously resolve CSV in the quark Mellin moments. These results reduce the NuTeV anomaly from 3σ to 2σ and could improve the sensitivity of Standard Model tests such as the PVDIS program at Jefferson Laboratory. The same lattice QCD studies

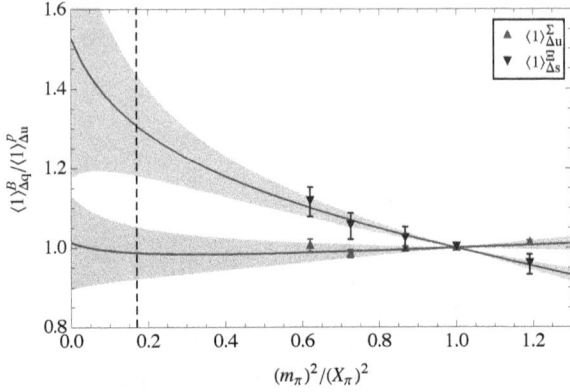

Fig. 4. Ratio of doubly-represented quark spin fractions in the octet baryons, taken from Ref. 18. X_π is the singlet quark mass.

show that the fraction of baryon spin carried by the quarks is structure-dependent, rather than universal across the baryon octet.

Acknowledgments

This work was supported by the University of Adelaide and the Australian Research Council through through the ARC Centre of Excellence for Particle Physics at the Terascale and grants FL0992247 (AWT), DP110101265 (RDY) and FT120100821 (RDY).

References

1. J. T. Londergan, J. C. Peng and A. W. Thomas, Rev. Mod. Phys. **82** (2010) 2009 [arXiv:0907.2352 [hep-ph]].
2. J. T. Londergan and A. W. Thomas, Prog. Part. Nucl. Phys. **41** (1998) 49.
3. G. A. Miller, A. K. Opper and E. J. Stephenson, Ann. Rev. Nucl. Part. Sci. **56** (2006) 253 [nucl-ex/0602021].
4. A. D. Martin *et al.*, Eur. Phys. J. C **35** (2004) 325 [hep-ph/0308087].
5. X. Zheng, P. Reimer, K. Paschke *et al.*, *Jefferson Lab Experiment E1207102*.
6. C. Y. Prescott *et al.*, Phys. Lett. B **84** (1979) 524.
7. T. Hobbs and W. Melnitchouk, Phys. Rev. D **77** (2008) 114023.
8. T. Hobbs, AIP Conf. Proc. **1369** (2011) 51 [arXiv:1102.1106 [hep-ph]].
9. S. Mantry, M. J. Ramsey-Musolf and G. F. Sacco, Phys. Rev. C **82** (2010) 065205 [arXiv:1004.3307 [hep-ph]].
10. A. Kurylov, M. J. Ramsey-Musolf and S. Su, Phys. Lett. B **582** (2004) 222.
11. P. E. Shanahan, A. W. Thomas and R. D. Young, Phys. Rev. D **87** (2013) 094515 [arXiv:1303.4806 [nucl-th]].

12. R. Horsley *et al.*, Phys. Rev. D **83** (2011) 051501 [arXiv:1012.0215 [hep-lat]].
13. I. C. Cloet *et al.*, Phys. Lett. B **714** (2012) 97 [arXiv:1204.3492 [hep-lat]].
14. G. P. Zeller *et al.* [NuTeV Collaboration], Phys. Rev. Lett. **88** (2002) 091802 [Erratum-ibid. **90** (2003) 239902] [hep-ex/0110059].
15. W. Bentz, I. C. Cloet, J. T. Londergan and A. W. Thomas, Phys. Lett. B **693** (2010) 462 [arXiv:0908.3198 [nucl-th]].
16. M. G. Alekseev *et al.* [COMPASS Collaboration], Phys. Lett. B **690** (2010) 466 [arXiv:1001.4654 [hep-ex]].
17. A. Deshpande, R. Milner, R. Venugopalan and W. Vogelsang, Ann. Rev. Nucl. Part. Sci. **55** (2005) 165 [hep-ph/0506148].
18. P. E. Shanahan *et al.*, Phys. Rev. Lett. **110** (2013) 202001.
19. P. E. Shanahan, A. W. Thomas and R. D. Young, Phys. Lett. B **718** (2013) 1148.
20. J. Beringer *et al.* [Particle Data Group Collaboration], Phys. Rev. D **86** (2012) 010001.
21. J. Gasser and H. Leutwyler, Phys. Rept. **87** (1982) 77.
22. A. Walker-Loud, C. E. Carlson and G. A. Miller, Phys. Rev. Lett. **108** (2012) 232301 [arXiv:1203.0254 [nucl-th]].
23. S. Aoki *et al.* [PACS-CS Collaboration], Phys. Rev. D **79** (2009) 034503.
24. W. Bietenholz *et al.*, Phys. Rev. D **84** (2011) 054509.
25. H. Leutwyler, Phys. Lett. B **378** (1996) 313 [hep-ph/9602366].
26. G. Colangelo *et al.*, Eur. Phys. J. C **71** (2011) 1695.
27. S. Borsanyi *et al.*, arXiv:1306.2287 [hep-lat].
28. R. Horsley *et al.* [QCDSF and UKQCD Collaborations], Phys. Rev. D **86** (2012) 114511 [arXiv:1206.3156 [hep-lat]].
29. G. M. de Divitiis *et al.*, JHEP **1204** (2012) 124 [arXiv:1110.6294 [hep-lat]].
30. T. Blum *et al.*, Phys. Rev. D **82** (2010) 094508 [arXiv:1006.1311 [hep-lat]].
31. S. R. Beane, K. Orginos and M. J. Savage, Nucl. Phys. B **768** (2007) 38.
32. E. N. Rodionov, A. W. Thomas and J. T. Londergan, Mod. Phys. Lett. A **9** (1994) 1799.
33. J. T. Londergan and A. W. Thomas, Phys. Rev. D **67** (2003) 111901.
34. D. Wang *et al.* [Jefferson Lab Hall A Collaboration], Phys. Rev. Lett. **111** (2013) 082501 [arXiv:1304.7741 [nucl-ex]].
35. F. Myhrer and A. W. Thomas, Phys. Rev. D **38** (1988) 1633.
36. F. Myhrer and A. W. Thomas, Phys. Lett. B **663** (2008) 302.
37. A. W. Schreiber and A. W. Thomas, Phys. Lett. B **215** (1988) 141.

Chirality in atomic nuclei: 2013

S. Q. Zhang*, Q. B. Chen

*State Key Laboratory of Nuclear Physics and Technology, School of Physics,
Peking University, Beijing 100871, China*
E-mail: sqzhang@pku.edu.cn

J. Meng[†]

*State Key Laboratory of Nuclear Physics and Technology, School of Physics,
Peking University, Beijing 100871, China*
*School of Physics and Nuclear Energy Engineering, Beihang University,
Beijing 100191, China*
Department of Physics, University of Stellenbosch, Stellenbosch, South Africa
[†]*E-mail: mengj@pku.edu.cn*

Progresses of the chirality in atomic nuclei are reviewed, in particular, the recently proposed collective Hamiltonian based on tilted axis cranking approach to describe chiral vibration and rotation modes, and the experimental achievements for chirality and multiple chiral doublets, i.e., in ^{106}Ag, ^{133}Ce, and ^{103}Rh. The first experimental evidences of multiple chiral doublet bands with distinct and identical configuration found in ^{133}Ce and ^{103}Rh are discussed in detail.

Keywords: Chirality; Band crossing; Collective Hamiltonian; Multiple chiral doublet bands.

1. Historical Review

Spontaneous chiral-symmetry breaking is a phenomenon of general interest, e.g., the chiral symmetry of the gauge theory in particle physics and the geometric properties of certain molecules in chemistry. Frauendorf and Meng originally suggested the existence of this phenomenon in rotating triaxial odd-odd nuclei in 1997.[1] The two odd nucleons align their angular momentum vectors along the short and long axes of the nucleus if they have a particle- and hole-like nature, respectively, while the collective rotation vector aligns along the intermediate axis. A left- and right-handed system will be generated in the intrinsic frame depending upon which side of the short-long plane the rotation vector projects from. In the laboratory frame, the restoration of spontaneous chiral-symmetry breaking manifests itself as a pair of degenerate $\Delta I = 1$ bands with the same parity, i.e., the chiral doublet bands.

Theoretically, the chiral doublets bands have been successfully described by a variety of formulations viz. the triaxial particle rotor model (PRM),[1–4] the interacting boson fermion-fermion model (IBFFM),[5–7] the tilted axis cranking (TAC) model with shell correction (SCTAC),[8] Skyrme-Hartree-Fock,[9,10] and the random phase approximation.[11,12] Recently, based on TAC with single-j shell, a new collective Hamiltonian to describe the chiral rotation and vibration was proposed and applied to the system with one $h_{11/2}$ proton particle and one $h_{11/2}$ neutron hole coupled to a triaxial rotor.[13]

Experimentally, the evidences of chiral doublet bands have been extensively reported in the $A \approx 80$, 100, 130 and 190 mass regions.[14–31] However, it should be pointed out that the lifetime measurements which are essential to extract the absolute electromagnetic transition probabilities are still rare for the candidate chiral bands. Examples in $A \sim 130$ mass region include ^{134}Pr,[5] ^{128}Cs,[23] ^{135}Nd,[11] ^{126}Cs.[32] In $A \sim 100$ mass region, the lifetime measurements have been performed for ^{103,104}Rh,[33] ^{102}Rh[34] and ^{106}Ag.[35,36]

Although the chiral partner bands have energies close to each other, it is rare to observe a crossing between them. Therefore crossing of candidate chiral bands is important as a test of the onset of static chirality. The most famous examples of such a crossing are in ^{134}Pr[37] and ^{106}Ag.[38] Recently, a thorough spectroscopic and theoretical investigation of the negative-parity bands in ^{106}Ag was performed and the origin of crossing was clarified.[35]

It was demonstrated by Meng et al.,[39–42] based on adiabatic and configuration-fixed constrained triaxial covariant density functional theory (CDFT) calculations, that it is possible to have multiple pairs of chiral doublet bands in a single nucleus, and the acronym MχD was introduced for this phenomenon.[39] In 2013, the first experimental evidence for the MχD in ^{133}Ce[43] was reported. Subsequently, the evidence of MχD was also suggested in ^{107}Ag.[44]

It is interesting to study the robustness of chiral geometry against the increase of the intrinsic excitation energy, i.e., whether the chiral geometry is sustained or not in the higher-lying bands of a certain chiral configuration.

Indeed, very recent model calculations predicted multiple chiral doublet bands which belong to the same configuration.[45–47] Recently, the first experimental evidence for such a new type of MχD was reported in ^{103}Rh.[48]

In this invited talk, we will briefly review the recent progresses of nuclear chirality including the collective Hamiltonian for chiral modes,[13] the resolution of chiral conundrum of crossing bands in ^{106}Ag,[35] the first experimental evidences for MχD bands with the distinct configurations in ^{133}Ce and the identical configuration in ^{103}Rh.[43,48]

2. Theoretical Progress: Collective Hamiltonian for Chiral Modes

As discussed in Sec. 1, there are various versions of TAC approaches[8–10] being employed to describe the chiral doublet bands. Based on mean field approximation, TAC approach can exhibit the orientation of the density distribution relative to the angular momentum vector. However, the description of quantum tunneling between chiral partners is beyond the scope of mean field approximation, thus only the yrast sequence can be obtained in TAC. To go beyond mean field approximation to yield the chiral partners, one can incorporate the quantum correlations by means of RPA[11,12] or collective Hamiltonian.[13] By taking into account the quantum fluctuation along the collective degree of freedom, the collective Hamiltonian goes beyond the mean field approximation and restores the broken symmetry.[13]

The detailed theoretical framework of collective Hamiltonian based on the TAC approach has been formulated in Ref. 13. Taking the azimuth angle of nucleus orientation φ as the collective variable, the quantized form of the collective Hamiltonian is written as the sum of kinetic and potential terms

$$
\begin{aligned}
\hat{H}_{\text{coll}} &= \hat{T}_{\text{kin}}(\varphi) + V(\varphi) \\
&= -\frac{\hbar^2}{2\sqrt{B(\varphi)}} \frac{\partial}{\partial\varphi} \frac{1}{\sqrt{B(\varphi)}} \frac{\partial}{\partial\varphi} + V(\varphi),
\end{aligned}
\tag{1}
$$

in which both the collective potential $V(\varphi)$ and mass parameter $B(\varphi)$ are obtained by TAC solutions. Solving this collective Hamiltonian on the basis states with appropriate boundary condition, e.g., box boundary condition, the collective levels and corresponding wave functions can be obtained.

In Fig. 1, the energy spectra of the doublet bands obtained from the collective Hamiltonian are compared with the exact solutions by PRM calculations for a system with one $h_{11/2}$ proton particle and one $h_{11/2}$ neutron hole coupled to a rigid rotor with $\gamma = -30°$. In the inset, the spin $I(\omega)$ obtained from TAC in comparison with PRM is also shown. The good agreement between the TAC and PRM for the spin is clearly seen. Furthermore, apart from the agreement of collective Hamiltonian and PRM

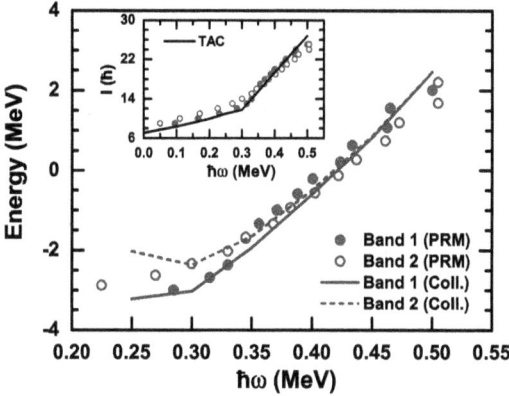

Fig. 1. (Color online) The energy spectra of the doublet bands obtained from the collective Hamiltonian in comparison with the exact solutions by the PRM. Inset: The spin $I(\omega)$ obtained from TAC in comparison with PRM. A similar $I(\omega)$ plot has been given in Ref. 1. Taken from Ref. 13.

results for the yrast band, the partner band of PRM can also be reasonably reproduced by the collective Hamiltonian.

For the collective Hamiltonian results, the energy differences between the doublet bands become smaller with the increase of the cranking frequency. For the PRM results, however, the doublet bands become closer up to $\hbar\omega \sim 0.35$ MeV and the energy differences between the doublet bands continue to increase for the higher cranking frequency. As demonstrated both in PRM[4,49,50] and TAC+RPA[12] investigations, the doublet bands will attain a second chiral vibration character, which is not taken into account by the present collective Hamiltonian investigation.

The success of the collective Hamiltonian here guarantees its application for realistic TAC calculations. In addition, it is mentioned that the collective Hamiltonian has been applied to investigate another novel rotational mode in triaxial nucleus — wobbling motion.[51] For the details, see Refs. 13 and 51.

3. Experimental Achievement I: Chiral Conundrum in ^{106}Ag

The chiral doublet bands at low spin are usually energetically separated and understood as a chiral vibrational mode.[4,11] With increasing spin, they approach each other and become approximately degenerate after a critical spin, forming thus the static chiral mode — a consequence of quantum tunneling between the left- and right-handed solutions.[4,11] Therefore, the appearance of crossing bands has drawn lots of attentions. In particular the nucleus ^{106}Ag, as one of only two known examples of candidates which actually cross, has been investigated by many groups.[35,36,38,52-54] In a recent work,[35] the high spin states in ^{106}Ag were populated using ^{96}Zr(^{14}N, $4n$) reaction with the γ-detector array AFRODITE at iThemba LABS. The level scheme of ^{106}Ag has been extended, and three negative-parity bands, labeled as bands 1, 2, and 3, have been observed to high spins. In particular, lifetimes have been determined for these bands using the Doppler-shift attenuation method.

The bands 1 and 2 cross each other around $I = 14\hbar$. Joshi et al.[38] have assigned $\pi g_{9/2}^{-1} \otimes \nu h_{11/2}$ configurations to them, while Ma et al.[53] assigned a $\pi g_{9/2}^{-1} \otimes \nu h_{11/2}^3$ configuration to band 2. An inspection of the aligned angular momenta reveals that band 1 has $\approx 6\hbar$ while band 2, as well as band 3, have $\approx 10\hbar$ as can be seen in Fig. 2. The different alignments were explained by Joshi et al.[38] as a result of aplanar and planar rotations for the partner bands, but an alternative of different configuration assignments is necessary by performing an empirical quasiparticle analysis.[56]

As can be seen in Fig. 2(a), the configuration $\pi g_{9/2}^{-1} \otimes \nu h_{11/2}$ assignment to band 1 in ^{106}Ag is confirmed below 0.5 MeV. Above 0.5 MeV, the alignment of band 1 indicates an onset of band crossing. For band 2, a good agreement is found for the alignment if a four-quasiparticle config-

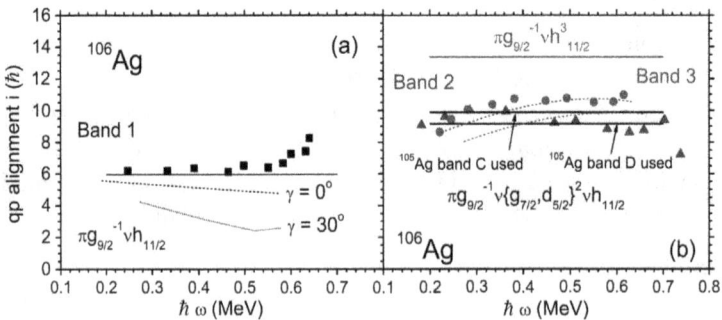

Fig. 2. (Color online) Quasiparticle alignment i as function of rotational frequency for (a) band 1 and (b) bands 2 and 3 in ^{106}Ag. The experimental points are shown as full symbols. The alignments deduced from the neighboring nuclei are indicated as straight lines. Results of theoretical calculations are displayed as dashed and dotted lines. The Harris parameters used are $\mathcal{J}_0 = 8.9 \ \hbar^2/\text{MeV}$ and $\mathcal{J}_1 = 15.7 \ \hbar^4/\text{MeV}^3$.[55] Taken from Ref. 35.

uration $\pi g_{9/2}^{-1} \otimes \nu\{g_{7/2}, d_{5/2}\}^2 \nu h_{11/2}$ is assigned to the band. In Fig. 2(b) the alignments of bands 2 and 3 in ^{106}Ag are compared with those deduced from neighboring odd-mass nuclei, using bands C and D in ^{105}Ag of $\pi g_{9/2}^{-1} \otimes \nu\{g_{7/2}, d_{5/2}\} \nu h_{11/2}$ configuration.[57] As can be seen, not only the aligned angular momenta of band 2 in ^{106}Ag, but also those of band 3 are quite well reproduced. The alternative $\pi g_{9/2}^{-1} \otimes \nu h_{11/2}^3$ configuration assignment for these bands, proposed by Ma et $al.$,[53] can be excluded because its alignment is $\approx 3\hbar$ larger than those of bands 2 and 3 (cf. Fig. 2(b)) which is a significant difference considering that in the neighboring nuclei the alignment relations are fulfilled to within 1-2$\hbar$.[55]

The excitation energies, $B(M1)$ and $B(E2)$ values, as well as $B(M1)/B(E2)$ ratios have been compared with results of PRM.[2–4] It is found that the PRM can reproduce the data well. In addition, the scenario proposed by Joshi et $al.$,[38] that bands 1 and 2 have $\pi g_{9/2}^{-1} \otimes \nu h_{11/2}$ configurations of triaxial and prolate shape, can be once again tested in the PRM against the experimental alignments. Calculations have been carried out for $\gamma = 0°$ and $30°$ shown as dotted and dashed curves, respectively, in Fig. 2(a). The $\gamma = 0°$ curve reproduces the experimental alignments reasonably well while a reduction in aligned angular momentum is predicted for $\gamma = 30°$. Neither curve is in agreement with the higher aligned angular momenta observed for bands 2 and 3. For bands 2 and 3, the calculated aligned angular momenta with four-quasiparticle configuration $\pi g_{9/2}^{-1} \otimes \nu\{g_{7/2}, d_{5/2}\} \nu h_{11/2}$ are in good agreement with experiment, as shown in Fig. 2(b).

From the investigations, it is concluded that the three close-lying negative-parity bands in ^{106}Ag are a two-quasiparticle high-K band and a pair of four-quasiparticle bands. It is clarified that the crossing between bands 1 and 2 is caused by configurations of different alignment.

For the details, see Ref. 35.

4. Experimental Achievement II: Multiple Chiral Doublet Bands - MχD

The phenomenon of MχD, i.e., more than one pair of chiral doublet bands exist in a single nucleus, was firstly predicted based on triaxial CDFT in ^{106}Rh,[39] where each pair of chiral doublet bands differ from each other in their triaxial deformations and multiparticle configurations. This phenomenon represents an important confirmation of triaxial shape coexistence in nuclei. Subsequently, it is pointed out that the phenomenon of MχD may also exist with the same particle-hole configuration, i.e., not only the yrast and yrare bands but also two higher excited bands might be chiral partners.[45–47] In the following, the corresponding experimental evidences of the two kinds of MχD observed in ^{133}Ce[43] and ^{103}Rh[48] will be briefly introduced. For the details, see Refs. 43 and 48.

4.1. *MχD with distinct configurations:* ^{133}*Ce*

After the prediction of MχD in the nuclei,[39] more and more efforts have been made to search for the corresponding experimental evidences. Using the ATLAS facility at the Argonne National Laboratory, high-spin states in ^{133}Ce were populated following the ^{116}Cd(^{22}Ne, $5n$) reaction. A more complete level scheme was obtained, of which two distinct sets of doublet bands (respectively labeled as bands 5 and 6 and bands 2 and 3) were identified.[43]

Calculations based on a combination of the constrained triaxial CDFT[39–42] and the triaxial PRM[2–4] were performed to investigate the nature of the observed pairs of doublet bands in ^{133}Ce. In Fig. 3, the energy

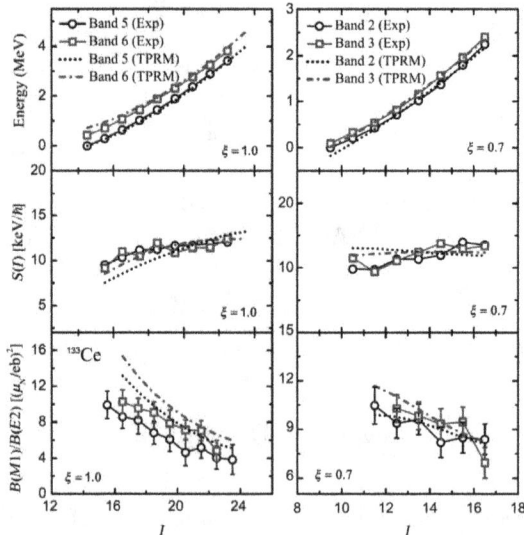

Fig. 3. (Color online) Experimental excitation energies, $S(I)$ parameters, and $B(M1)/B(E2)$ ratios for the negative-parity chiral doublet (left panels) and positive-parity chiral doublet (right panels) in 133Ce. Also shown are results of PRM calculations with the indicated attenuation factors ξ (see text). Taken from Ref. 43.

spectra, $S(I)$ parameters, and $B(M1)/B(E2)$ calculated by PRM are in comparison with the experimental values for the negative parity bands 5 and 6 as well as the positive parity bands 2 and 3. For band 5 and 6, they are based on a $\pi(1h_{11/2})^2 \otimes \nu(1h_{11/2})^{-1}$ configuration. The theoretical results show an impressive agreement with the data. The energy separation between the two bands is about 400 keV at $I = 29/2$, and remains relatively constant over an extended spin range. The energy variation between bands 5 and 6 similarly suggests a form of chiral vibration, a tunneling between the left- and right-handed configurations, such that bands 5 and 6 are associated with the zero- and one-phonon state, respectively. Furthermore, the calculated staggering parameter $S(I)$ is seen to vary smoothly with spin. This is expected since the Coriolis interaction is substantially reduced for a three-dimensional coupling of angular momentum vectors in a chiral geometry. The calculated electromagnetic transition probability ratio in the PRM has no obvious odd-even staggering of the $B(M1)/B(E2)$ values, although a small effect is apparent experimentally. This further supports the interpretation in terms of a chiral vibration between bands 5 and 6.[49]

The positive parity bands 2 and 3 are associated with the configuration $\pi[(1g_{7/2})^{-1}(1h_{11/2})^1] \otimes \nu(1h_{11/2})^{-1}$. When comparing the experimental energies with the PRM calculations, a Coriolis attenuation factor, $\xi = 0.7$, was employed considering the fact that the configuration contains a low-j orbital $\pi(1g_{7/2})$ which has large admixtures with other low-j orbitals. The energy separation between bands 2 and 3 was found to be nearly constant at ~ 100 keV, which, combined with the spin-independent $S(I)$ parameter, leads to the interpretation of these bands being chiral partners as well. The similar behavior of $B(M1)/B(E2)$ ratios is also clearly evident. Similar to bands 5 and 6, there is no perceptible staggering in the calculated $B(M1)/B(E2)$ ratios, but such an effect is apparent experimentally. The $B(M1)/B(E2)$ ratio depends sensitively on the details of the transition from the vibrational to the tunneling regime,[49] which may account for the deviations of the PRM calculation from experiment.

4.2. $M\chi D$ with identical configuration: ^{103}Rh

Medium- and high-spin states of ^{103}Rh were populated using the ^{96}Zr(^{11}B,4n) reaction at a beam energy of 40 MeV.[48] The beam, provided by the 88-inch Cyclotron of the Lawrence Berkeley National Laboratory, impinged upon an enriched 500 μg/cm^2 thick self-supporting Zr foil. The emitted γ-rays were detected by the Gammasphere spectrometer. A more complete level scheme of ^{103}Rh, including three positive parity and five negative parity bands, was constructed using the observed coincidence relations and relative intensities of the gamma transitions and based on the formerly reported states.[29,58]

In order to understand the nature of the observed band structure in ^{103}Rh, the adiabatic and configuration-fixed constrained CDFT calculations[39] were performed to search for the possible configurations and deformations. Subsequently, the configurations and deformations were further confirmed and reexamined by tilted axis cranking CDFT (TAC-CDFT) calculations[59–62] determining the energy spectra, Routhians, spin-frequency relations, deformations, and alignments. Finally, with the obtained configurations and deformations, quantum PRM[2–4] calculations were performed to study the energy spectra and $B(M1)/B(E2)$ ratios for both the positive- and the negative-parity bands.

The obtained energy spectra by PRM are in comparison with data in Fig. 4. For bands 1 and 2, the PRM results excellently agree with the data. These two bands are separated by ~500 keV at $I = 29/2\ \hbar$. They approach each other with increasing spin and the separation finally goes to ~ 360

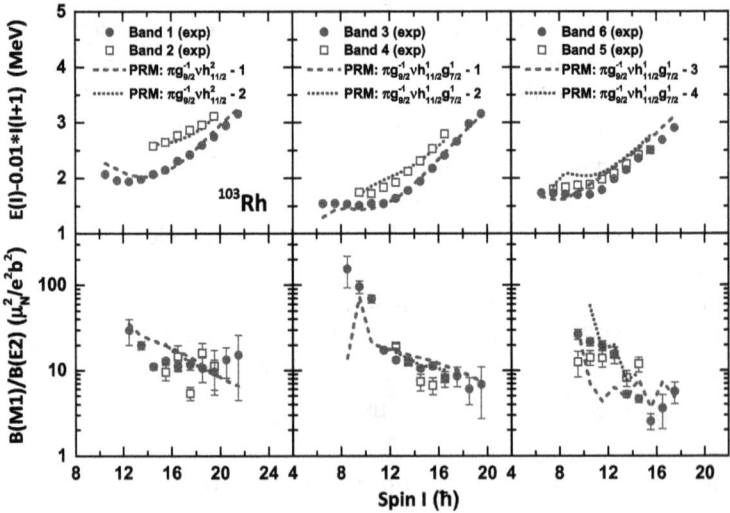

Fig. 4. (Color online) Experimental excitation energies and $B(M1)/B(E2)$ ratios for the positive-parity chiral bands 1-2 (left panels) and negative-parity multiple chiral bands 3-6 (middle and right panels) in ^{103}Rh together with the results of triaxial particle rotor model. The number following the configuration label of the theoretical curve corresponds to the energy ordering of the calculated band with the given configuration. Taken from Ref. 48.

keV at $I = 39/2$. The $B(M1)/B(E2)$ values of bands 1 and 2 are similar. The observation that the experimental $B(M1)/B(E2)$ values for bands 1 and 2 do not fall off as quickly with spin as the theoretical values comes from the frozen rotor assumption adopted in PRM. As discussed in Ref. 50, the chiral bands with positive parity change from chiral vibration to nearly static chirality at spin $I = 37/2$ and back to another type of chiral vibration at higher spins. Such a conclusion is still held here for the positive-parity doublet.

Special attentions were made to the four negative-parity bands 3-6, where the configuration $\pi(1g_{9/2})^{-1} \otimes \nu(1h_{11/2})^1(1g_{7/2})^1$ is adopted to describe the experimental data.[48] The four calculated bands form two chiral doublets, of which the first one fits the experimental band-pair 3 and 4, while the second doublet can also reasonably reproduce the trend of bands 6 and 5. The calculated energies for bands 5 and 6 are higher than the experimental values about 200 keV, which might be ascribed to that the complex correlations are not fully taken into account in the PRM calculations with single-j shell Hamiltonian. The corresponding calculated electromagnetic transition probabilities, shown in Fig. 4, are also able to reproduce the data reasonably. The weak odd-even $B(M1)/B(E2)$ staggering for bands 3 and 4 is consistent with the case of chiral vibration as discussed in Ref. 49. For bands 5 and 6, the $B(M1)/B(E2)$ values show a staggering at $I = 15.5\ \hbar$, which is also reproduced by the PRM.

In contrast with the multiple chiral doublets predicted in Ref. 39 and experimentally reported in ^{133}Ce,[43] the observed MχD in the negative-parity bands of ^{103}Rh is built from the first and second doublets of the same configuration. Observation of MχD with the same configuration shows that the chiral geometry in nuclei can be robust against the increase of the intrinsic excitation energy.

5. Summary

The recent progresses of the chirality and MχD in atomic nuclei are briefly reviewed for both theoretical and experimental sides. The theoretical side is focused on the introduction of the collective Hamiltonian based on tilted axis cranking approach proposed to describe chiral vibration and rotation modes.[13] The experimental achievements reviewed here include the newly reported results in ^{106}Ag, ^{133}Ce, and ^{103}Rh, i.e., the band crossing conundrum in ^{106}Ag,[35] the first experimental evidences of MχD in ^{133}Ce,[43] and the MχD with identical configuration in ^{103}Rh.[48]

Acknowledgments

This work was supported in part by the Major State 973 Program of China (Grant No. 2013CB834400), the National Natural Science Foundation of China (Grants No. 11175002, No. 11335002, No. 11375015, No. 11345004), Research Fund for the Doctoral Program of Higher Education (Grant No. 20110001110087).

References

1. S. Frauendorf and J. Meng, *Nucl. Phys. A* **617**, p. 131 (1997).
2. J. Peng, J. Meng and S. Q. Zhang, *Phys. Rev. C* **68**, p. 044324 (2003).
3. S. Q. Zhang, B. Qi, S. Y. Wang and J. Meng, *Phys. Rev. C* **75**, p. 044307 (2007).
4. B. Qi, S. Q. Zhang, J. Meng, S. Y. Wang and S. Frauendorf, *Phys. Lett. B* **675**, p. 175 (2009).
5. D. Tonev, G. de Angelis, P. Petkov, A. Dewald, S. Brant, S. Frauendorf, D. L. Balabanski, P. Pejovic, D. Bazzacco, P. Bednarczyk, F. Camera, A. Fitzler, A. Gadea, S. Lenzi, S. Lunardi, N. Marginean, O. Moller, D. R. Napoli, A. Paleni, C. M. Petrache, G. Prete, K. O. Zell, Y. H. Zhang, J. Y. Zhang, Q. Zhong and D. Curien, *Phys. Rev. Lett.* **96**, p. 052501 (2006).
6. D. Tonev, G. de Angelis, S. Brant, S. Frauendorf, P. Petkov, A. Dewald, F. Doenau, D. L. Balabanski, Q. Zhong, P. Pejovic, D. Bazzacco, P. Bednarczyk, F. Camera, D. Curien, F. Della Vedova, A. Fitzler, A. Gadea, G. Lo Bianco, S. Lenzi, S. Lunardi, N. Marginean, O. Moeller, D. R. Napoli, R. Orlandi, E. Sahin, A. Saltarelli, J. V. Dobon, K. O. Zell, J.-y. Zhang and Y. H. Zhang, *Phys. Rev. C* **76**, p. 044313 (2007).
7. S. Brant, D. Tonev, G. de Angelis and A. Ventura, *Phys. Rev. C* **78**, p. 034301 (2008).
8. V. I. Dimitrov, S. Frauendorf and F. Donau, *Phys. Rev. Lett.* **84**, 5732 (2000).
9. P. Olbratowski, J. Dobaczewski, J. Dudek and W. Plociennik, *Phys. Rev. Lett.* **93**, p. 052501 (2004).
10. P. Olbratowski, J. Dobaczewski and J. Dudek, *Phys. Rev. C* **73**, p. 054308 (2006).
11. S. Mukhopadhyay, D. Almehed, U. Garg, S. Frauendorf, T. Li, P. V. M. Rao, X. Wang, S. S. Ghugre, M. P. Carpenter, S. Gros, A. Hecht, R. V. F. Janssens, F. G. Kondev, T. Lauritsen, D. Seweryniak and S. Zhu, *Phys. Rev. Lett.* **99**, p. 172501 (2007).
12. D. Almehed, F. Dönau and S. Frauendorf, *Phys. Rev. C* **83**, p. 054308 (2011).
13. Q. B. Chen, S. Q. Zhang, P. W. Zhao, R. V. Jolos and J. Meng, *Phys. Rev. C* **87**, p. 024314 (2013).
14. K. Starosta, T. Koike, C. J. Chiara, D. B. Fossan, D. R. LaFosse, A. A. Hecht, C. W. Beausang, M. A. Caprio, J. R. Cooper, R. Krücken, J. R. Novak, N. V. Zamfir, K. E. Zyromski, D. J. Hartley, D. L. Balabanski, J.-y. Zhang, S. Frauendorf and V. I. Dimitrov, *Phys. Rev. Lett.* **86**, 971 (2001).
15. A. A. Hecht, C. W. Beausang, K. E. Zyromski, D. L. Balabanski, C. J.

Barton, M. A. Caprio, R. F. Casten, J. R. Cooper, D. J. Hartley, R. Krucken, D. Meyer, H. Newman, J. R. Novak, E. S. Paul, N. Pietralla, A. Wolf, N. V. Zamfir, J. Y. Zhang and F. Donau, *Phys. Rev. C* **63**, p. 051302 (2001).

16. D. J. Hartley, L. L. Riedinger, M. A. Riley, D. L. Balabanski, F. G. Kondev, R. W. Laird, J. Pfohl, D. E. Archer, T. B. Brown, R. M. Clark, M. Devlin, P. Fallon, I. M. Hibbert, D. T. Joss, D. R. LaFosse, P. J. Nolan, N. J. O'Brien, E. S. Paul, D. G. Sarantites, R. K. Sheline, S. L. Shepherd, J. Simpson, R. Wadsworth, J. Y. Zhang, P. B. Semmes and F. Donau, *Phys. Rev. C* **64**, p. 031304 (2001).

17. T. Koike, K. Starosta, C. J. Chiara, D. B. Fossan and D. R. LaFosse, *Phys. Rev. C* **63**, p. 061304 (2001).

18. R. A. Bark, A. M. Baxter, A. P. Byrne, G. D. Dracoulis, T. Kibedi, T. R. McGoram and S. M. Mullins, *Nucl. Phys. A* **691**, 577 (2001).

19. K. Starosta, C. J. Chiara, D. B. Fossan, T. Koike, T. T. S. Kuo, D. R. LaFosse, S. G. Rohozinski, C. Droste, T. Morek and J. Srebrny, *Phys. Rev. C* **65**, p. 044328 (2002).

20. T. Koike, K. Starosta, C. J. Chiara, D. B. Fossan and D. R. LaFosse, *Phys. Rev. C* **67**, p. 044319 (2003).

21. G. Rainovski, E. S. Paul, H. J. Chantler, P. J. Nolan, D. G. Jenkins, R. Wadsworth, P. Raddon, A. Simons, D. B. Fossan, T. Koike, K. Starosta, C. Vaman, E. Farnea, A. Gadea, T. Kroll, R. Isocrate, G. de Angelis, D. Curien and V. I. Dimitrov, *Phys. Rev. C* **68**, p. 024318 (2003).

22. S. Zhu, U. Garg, B. K. Nayak, S. S. Ghugre, N. S. Pattabiraman, D. B. Fossan, T. Koike, K. Starosta, C. Vaman, R. V. F. Janssens, R. S. Chakrawarthy, M. Whitehead, A. O. Macchiavelli and S. Frauendorf, *Phys. Rev. Lett.* **91**, p. 132501 (2003).

23. E. Grodner, J. Srebrny, A. A. Pasternak, I. Zalewska, T. Morek, C. Droste, J. Mierzejewski, M. Kowalczyk, J. Kownacki, M. Kisielinski, S. G. Rohozinski, T. Koike, K. Starosta, A. Kordyasz, P. J. Napiorkowski, M. Wolinska-Cichocka, E. Ruchowska, W. Plociennik and J. Perkowski, *Phys. Rev. Lett.* **97**, p. 172501 (2006).

24. C. Vaman, D. B. Fossan, T. Koike, K. Starosta, I. Y. Lee and A. O. Macchiavelli, *Phys. Rev. Lett.* **92**, p. 032501 (2004).

25. P. Joshi, D. G. Jenkins, P. M. Raddon, A. J. Simons, R. Wadsworth, A. R. Wilkinson, D. B. Fossan, T. Koike, K. Starosta, C. Vaman, J. Timar, Z. Dombradi, A. Krasznahorkay, J. Molnar, D. Sohler, L. Zolnai, A. Algora, E. S. Paul, G. Rainovski, A. Gizon, J. Gizon, P. Bednarczyk, D. Curien, G. Duchene and J. N. Scheurer, *Phys. Lett. B* **595**, 135 (2004).

26. P. Joshi, A. R. Wilkinson, T. Koike, D. B. Fossan, S. Finnigan, E. S. Paul, P. M. Raddon, G. Rainovski, K. Starosta, A. J. Simons, C. Vaman and R. Wadsworth, *Eur. Phys. J. A* **24**, 23 (2005).

27. J. A. Alcántara-Núñez, J. R. B. Oliveira, E. W. Cybulska, N. H. Medina, M. N. Rao, R. V. Ribas, M. A. Rizzutto, W. A. Seale, F. Falla-Sotelo, K. T. Wiedemann, V. I. Dimitrov and S. Frauendorf, *Phys. Rev. C* **69**, p. 024317 (2004).

28. J. Timár, P. Joshi, K. Starosta, V. I. Dimitrov, D. B. Fossan, J. Molnar,

D. Sohler, R. Wadsworth, A. Algora, P. Bednarczyk, D. Curien, Z. Dombradi, G. Duchene, A. Gizon, J. Gizon, D. G. Jenkins, T. Koike, A. Krasznahorkay, E. S. Paul, P. M. Raddon, G. Rainovski, J. N. Scheurer, A. Simons, C. Vaman, A. R. Wilkinson, L. Zolnai and S. Frauendorf, *Phys. Lett. B* **598**, 178 (2004).

29. J. Timár, C. Vaman, K. Starosta, D. B. Fossan, T. Koike, D. Sohler, I. Y. Lee and A. O. Macchiavelli, *Phys. Rev. C* **73**, p. 011301 (2006).

30. D. L. Balabanski, M. Danchev, D. J. Hartley, L. L. Riedinger, O. Zeidan, J.-y. Zhang, C. J. Barton, C. W. Beausang, M. A. Caprio, R. F. Casten, J. R. Cooper, A. A. Hecht, R. Krücken, J. R. Novak, N. V. Zamfir and K. E. Zyromski, *Phys. Rev. C* **70**, p. 044305 (2004).

31. S. Y. Wang, B. Qi, L. Liu, S. Q. Zhang, H. Hua, X. Q. Li, Y. Y. Chen, L. H. Zhu, J. Meng, S. M. Wyngaardt, P. Papka, T. T. Ibrahim, R. A. Bark, P. Datta, E. A. Lawrie, J. J. Lawrie, S. N. T. Majola, P. L. Masiteng, S. M. Mullins, J. Gál, G. Kalinka, J. Molnár, B. Nyakó, J. Timár, K. Juhasz and R. Schwengner, *Physics Letters B* **703**, 40 (2011).

32. E. Grodner, I. Sankowska, T. Morek, S. G. Rohozinski, C. Droste, J. Srebrny, A. A. Pasternak, M. Kisielinski, M. Kowalczyk, J. Kownacki, J. Mierzejewski, A. Krol and K. Wrzosek, *Phys. Lett. B* **703**, 46 (2011).

33. T. Suzuki, G. Rainovski, T. Koike, T. Ahn, M. P. Carpenter, A. Costin, M. Danchev, A. Dewald, R. V. F. Janssens, P. Joshi, C. J. Lister, O. Moeller, N. Pietralla, T. Shinozuka, J. Timar, R. Wadsworth, C. Vaman and S. Zhu, *Phys. Rev. C* **78**, p. 031302 (2008).

34. D. Tonev, M. S. Yavahchova, N. Goutev, G. de Angelis, P. Petkov, R. K. Bhowmik, R. P. Singh, S. Muralithar, N. Madhavan, R. Kumar, M. Kumar Raju, J. Kaur, G. Mohanto, A. Singh, N. Kaur, R. Garg, A. Shukla, T. K. Marinov and S. Brant, *Phys. Rev. Lett.* **112**, p. 052501 (2014).

35. O. Lieder, E. M. Lieder, R. A. Bark, R. B. Chen, Q. Q. Zhang, S. J. Meng, A. Lawrie, E. J. Lawrie, J. P. Bvumbi, S. Y. Kheswa, N. S. Ntshangase, S. E. Madiba, T. L. Masiteng, P. M. Mullins, S. S. Murray, P. Papka, G. Roux, D. O. Shirinda, H. Zhang, Z. W. Zhao, P. P. Li, Z. J. Peng, B. Qi, Y. Wang, S. G. Xiao, Z. and C. Xu, *Phys. Rev. Lett.* **112**, p. 202502 (2014).

36. N. Rather, P. Datta, S. Chattopadhyay, S. Rajbanshi, A. Goswami, H. Bhat, G. A. Sheikh, J. S. Roy, R. Palit, S. Pal, S. Saha, J. Sethi, S. Biswas, P. Singh and C. Jain, H. *Phys. Rev. Lett.* **112**, p. 202503 (2014).

37. C. M. Petrache, G. B. Hagemann, I. Hamamoto and K. Starosta, *Phys. Rev. Lett.* **96**, p. 112502 (2006).

38. P. Joshi, M. P. Carpenter, D. B. Fossan, T. Koike, E. S. Paul, G. Rainovski, K. Starosta, C. Vaman and R. Wadsworth, *Phys. Rev. Lett.* **98**, p. 102501 (2007).

39. J. Meng, J. Peng, S. Q. Zhang and S.-G. Zhou, *Phys. Rev. C* **73**, p. 037303 (2006).

40. J. Peng, H. Sagawa, S. Q. Zhang, J. M. Yao, Y. Zhang and J. Meng, *Phys. Rev. C* **77**, p. 024309 (2008).

41. J. M. Yao, B. Qi, S. Q. Zhang, J. Peng, S. Y. Wang and J. Meng, *Phys. Rev. C* **79**, p. 067302 (2009).

42. J. Li, S. Q. Zhang and J. Meng, *Phys. Rev. C* **83**, p. 037301 (2011).

43. A. D. Ayangeakaa, U. Garg, M. D. Anthony, S. Frauendorf, J. T. Matta, B. K. Nayak, D. Patel, Q. B. Chen, S. Q. Zhang, P. W. Zhao, B. Qi, J. Meng, R. V. F. Janssens, M. P. Carpenter, C. J. Chiara, F. G. Kondev, T. Lauritsen, D. Seweryniak, S. Zhu, S. S. Ghugre and R. Palit, *Phys. Rev. Lett.* **110**, p. 172504 (2013).

44. B. Qi, H. Jia, N. B. Zhang, C. Liu and S. Y. Wang, *Phys. Rev. C* **88**, p. 027302 (2013).

45. C. Droste, S. G. Rohozinski, K. Starosta, L. Prochniak and E. Grodner, *Eur. Phys. J. A* **42**, 79 (2009).

46. Q. B. Chen, J. M. Yao, S. Q. Zhang and B. Qi, *Phys. Rev. C* **82**, p. 067302 (2010).

47. I. Hamamoto, *Phys. Rev. C* **88**, p. 024327 (2013).

48. I. Kuti, Q. B. Chen, J. Timár, D. Sohler, S. Q. Zhang, Z. H. Zhang, P. W. Zhao, J. Meng, K. Starosta, T. Koike, E. S. Paul, D. Fossan and C. Vaman, *Phys. Rev. Lett.* **113**, p. 032501(Jul 2014).

49. B. Qi, S. Q. Zhang, S. Y. Wang, J. M. Yao and J. Meng, *Phys. Rev. C* **79**, p. 041302(R) (2009).

50. B. Qi, S. Q. Zhang, S. Y. Wang, J. Meng and T. Koike, *Phys. Rev. C* **83**, p. 034303 (2011).

51. Q. B. Chen, S. Q. Zhang, P. W. Zhao and J. Meng, *arXiv:1407.3563 [nucl-th]* (2014).

52. C. Y. He, L. H. Zhu, X. G. Wu, S. X. Wen, G. S. Li, Y. Liu, Z. M. Wang, X. Q. Li, X. Z. Cui, H. B. Sun, R. G. Ma and C. X. Yang, *Phys. Rev. C* **81**, p. 057301(May 2010).

53. H. L. Ma, S. H. Yao, B. G. Dong, X. G. Wu, H. Q. Zhang and X. Z. Zhang, *Phys. Rev. C* **88**, p. 034322 (2013).

54. Y. Zheng, L. H. Zhu, X. G. Wu, C. Y. He, G. S. Li, X. Hao, B. B. Yu, S. H. Yao, B. Zhang, C. Xu, J. G. Wang and L. Gu, *Chin. Phys. Lett.* **31**, p. 62101 (2014).

55. H.-J. Keller, S. Frauendorf, U. Hagemann, L. Käubler, H. Prade and F. Stary, *Nuclear Physics A* **444**, 261 (1985).

56. R. Bengtsson and S. Frauendorf, *Nucl. Phys. A* **327**, 139 (1979).

57. J. Timár, T. Koike, N. Pietralla, G. Rainovski, D. Sohler, T. Ahn, G. Berek, A. Costin, K. Dusling, T. C. Li, E. S. Paul, K. Starosta and C. Vaman, *Phys. Rev. C* **76**, p. 024307 (2007).

58. H. Dejbakhsh, R. P. Schmitt and G. Mouchaty, *Phys. Rev. C* **37**, 621 (1988).

59. P. W. Zhao, S. Q. Zhang, J. Peng, H. Z. Liang, P. Ring and J. Meng, *Phys. Lett. B* **699**, 181 (2011).

60. P. W. Zhao, J. Peng, H. Z. Liang, P. Ring and J. Meng, *Phys. Rev. Lett.* **107**, p. 122501 (2011).

61. P. W. Zhao, J. Peng, H. Z. Liang, P. Ring and J. Meng, *Phys. Rev. C* **85**, p. 054310 (2012).

62. J. Meng, J. Peng, S. Q. Zhang and P. W. Zhao, *Front. Phys.* **8**, p. 55 (2013).